R W Dickey
University of Wisconsin-Madison

# Bifurcation problems in nonlinear elasticity

Pitman Publishing
LONDON · SAN FRANCISCO · MELBOURNE

PITMAN PUBLISHING LIMITED
39 Parker Street, London WC2B 5PB

FEARON–PITMAN INC.
6 Davis Drive, Belmont, California 94002, USA

*Associated Companies*
Copp Clark Ltd, Toronto
Pitman Publishing Co. SA (Pty) Ltd, Johannesburg
Pitman Publishing New Zealand Ltd, Wellington
Pitman Publishing Pty Ltd, Melbourne
Sir Isaac Pitman Ltd, Nairobi

AMS Subject Classifications: 73H05; 73K05, 73K10, 73K15

First published 1976
Reprinted 1977



Reproduced and printed by photolithography
in Great Britain at Biddles of Guildford

ISBN 0 273 00103 5

# Contents


| | | |
|---|---|---|
| 1 | Introduction | 1 |
| 2 | A static problem for the nonlinear string | 29 |
| 3 | A static problem for the nonlinear circular membrane | 42 |
| 4 | The rotating string | 56 |
| 5 | Existence of positive solutions | 76 |
| 6 | Bifurcation theory for second order ordinary differential equations. Application to the inextensible elastica | 87 |
| 7 | Buckling of the circular plate | 108 |

# 1 Introduction

In this set of lectures the problem of solving

(1.1) $F(u,\lambda) = 0$

will be discussed. In the most general context, F is a nonlinear transformation defined when $u \in s$ (s = some unspecified set of elements) and $\lambda$ a real parameter such that $a < \lambda < b$. The simplest situation occurs when (1.1) has a solution which is a single valued function of $\lambda$ so that the solution can be written $u = u(\lambda)$. Heuristically, this situation can be represented by a diagram as in fig. 1.1.

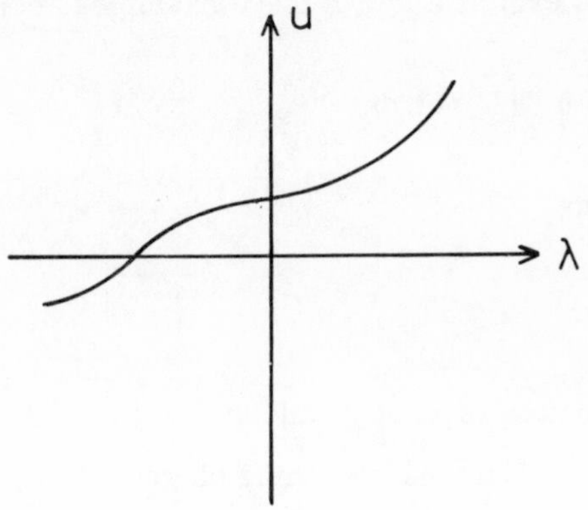


fig. 1.1

(Care must be exercised in the interpretation of fig. 1.1, since in general u would be a vector function and could not be accurately described by a single axis). A more typical situation arises when the solution of (1.1) is not a single valued function of $\lambda$. In this case much more complicated pictures are possible (cf. fig. 1.2)

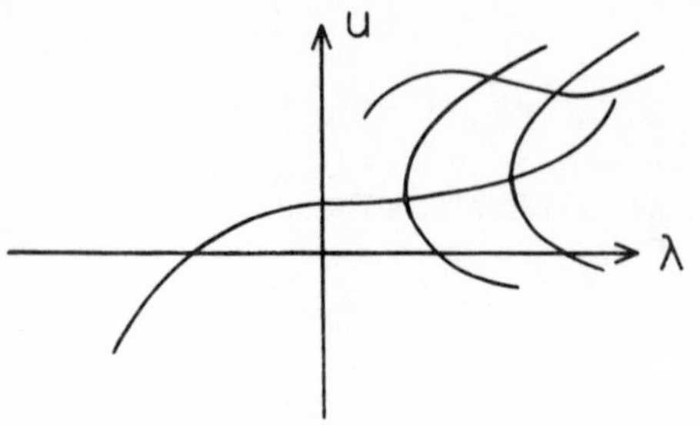


fig. 1.2

The object of bifurcation theory is to treat (1.1) when solutions are not single valued functions of $\lambda$.

Many examples of bifurcation phenomena occur in both differential and integral equations. In order to exhibit a few of these examples, consider the equation

(1.2) $w'' + \lambda w = 0 \quad (' = d/dx)$

with boundary conditions

(1.3) $w(0) = w(\pi) = 0$

The solution of this equation is $w \equiv 0$ unless $\lambda = m^2$. If $\lambda = m^2$ equ. (1.2) has a solution $w = A \sin mx$. Thus the branching diagram in fig. 1.3 is appropriate to (1.2).

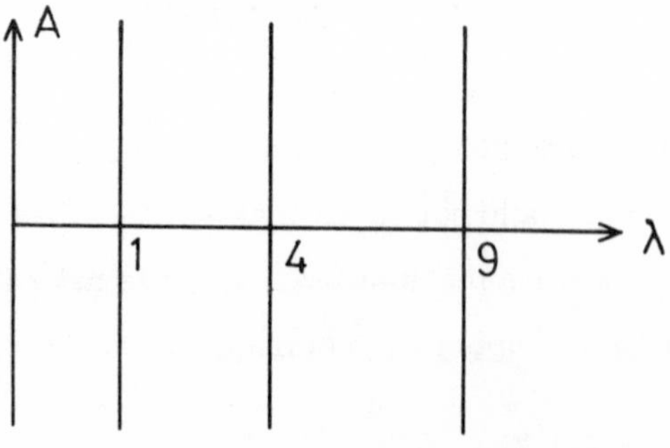


fig. 1.3

Other examples of this sort are furnished by the equations

$$(1.4) \qquad w'' + (\lambda - \frac{2}{\pi}\int_0^{\pi} w^2 dx)w = 0,$$

$$(1.5) \qquad (\frac{2}{\pi}\int_0^{\pi} w^2 dx)w'' + \lambda w = 0$$

with boundary conditions (1.3). It is easily verified that (1.4) and (1.5) have a solution $w = A \sin nx$ if $A$ satisfies the condition $A(-n^2 + (\lambda - A^2)) = 0$ in the case of equ. (1.4), and $A(-A^2n^2 + \lambda) = 0$ in the case of equ. (1.5). The corresponding branching diagrams are given in fig. 1.4.

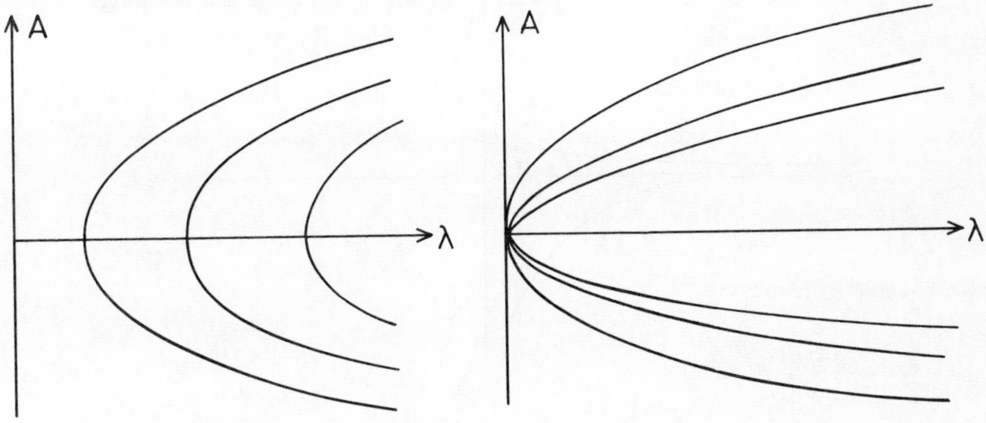


fig. 1.4

Another example is furnished by the integral equation

$$(1.6) \qquad \varphi = 1 + \lambda\int_0^1 \varphi^2 dx.$$

This equation is easily solved explicitly by simply squaring both sides of (1.6) and integrating from 0 to 1. The result is that

$$(1.7) \qquad \int_0^1 \varphi^2 dx = 1 + 2\lambda\int_0^1 \varphi^2 dx + \lambda^2(\int_0^1 \varphi^2 dx)^2$$

so that

$$(1.8) \qquad \int_0^1 \varphi^2 dx = \frac{1 - 2\lambda \pm \sqrt{(1 - 4\lambda)}}{2\lambda^2} .$$

Combining this result with (1.6), it follows that

$$(1.9) \qquad \varphi = \frac{1 \pm \sqrt{(1 - 4\lambda)}}{2\lambda} .$$

The resulting diagram is given in fig. 1.5.

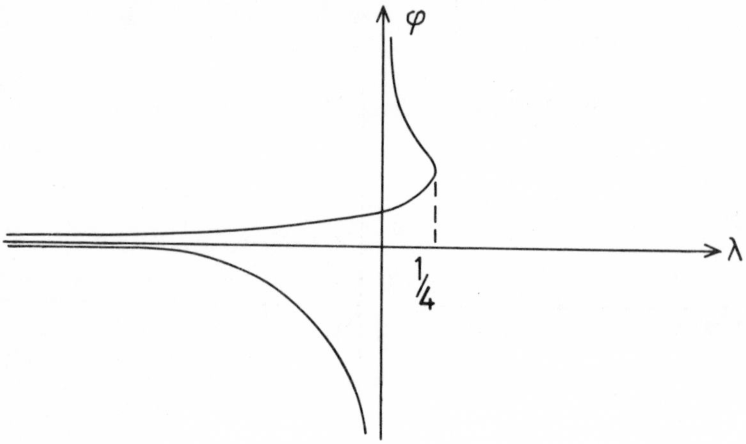


fig. 1.5

In view of the repeated occurrences of multiple 'branches' of solutions in differential and integral equations, it is to be expected that these play a significant role in the applications. In fact, this is the case - particularly in the various branches of continuum mechanics. In these lectures results pertaining to various problems arising in the applications will be described.

## Linear Theory, Function Spaces

As is to be expected, the most convenient place to begin a study of bifurcation theory is with linear transformations. For this purpose it is necessary to discuss the function spaces on which the transformations are to be defined.

Definition: A normed linear space is a linear space on which is defined a real valued function, called the norm, with the properties

1) $\| x \| \geq 0$ and $\| x \| = 0$ iff $x \equiv 0$,

2) $\| \alpha x \| = |\alpha| \, \| x \|$ ($\alpha$ a scalar),

3) $\| x_1 + x_2 \| \leq \| x_1 \| + \| x_2 \|$

Definition: A space is complete if every Cauchy sequence converges.

In explanation of this definition, it suffices to note that $(x_m)$ is a Cauchy sequence if $\lim_{n,m\to\infty} \| x_n - x_m \| = 0$. Thus a space is complete if to every Cauchy sequence there corresponds a vector $x$ such that $\lim_{m\to\infty} \| x_m - x \| = 0$. In this case it is said that $x_m$ converges to $x$ and is written $x_m \to x$.

Definition: An inner product space is a linear space on which is defined a complex valued function of two variables, called the inner product, with the properties

1) $\langle x,y \rangle = \overline{\langle y,x \rangle}$

2) $\langle \alpha x,y \rangle = \alpha \langle x,y \rangle$

3) $\langle x_1 + x_2,y \rangle = \langle x_1,y \rangle + \langle x_2,y \rangle$

4) $\langle x,x \rangle \geq 0$ and $\langle x,x \rangle = 0$ iff $x \equiv 0$.

It is easily verified that $\| x \| = \langle x,x \rangle^{\frac{1}{2}}$ is a norm in the sense defined above. The following lemma is easily proven.

Lemma (1.1) : (Schwartz Inequality).

(1.10) $|<x,y>| \leq \|x\| \|y\|$.

Definition: An inner product space which is complete under the inner product norm is a Hilbert space.

There are many examples of Hilbert and Banach spaces. For the purpose of these lectures, two of the most important are the Banach space $C(0,1) = \{f(x) |$ continuous or the interval $a \leq x \leq b\}$ with the norm $\|f\| = \max_{a \leq x \leq b} |f(x)|$, and the Hilbert space $\mathcal{L}_2(a,b)$, defined to be the completion under the norm $\|f\| = (\int_a^b |f(x)|^2 dx)^{\frac{1}{2}}$ of the functions which are continuous on $a \leq x \leq b$. The inner product defined on $\mathcal{L}_2(a,b)$ is simply $<f,g> = \int_a^b f(x)\bar{y}(x)dx$.

Definition: Two vectors $x$ and $y$ are orthogonal if $<x,y> = 0$.

Definition: A set of vectors $\{\phi_m\}$ are an orthonormal set if

$$< \phi_n,\phi_m > \delta_{nm} = \begin{cases} 0 n \neq m \\ 1 n = m. \end{cases}$$

Theorem 1.1 (Riemann-Lebesgue Lemma): If $\{\phi_n\}$ is an infinite orthonormal set and $f \epsilon H$ then

(1.11) $\lim_{n \to \infty} < f,\phi_n > 0.$

Proof: It follows from the definition of the inner product that

(1.12) $$< f - \sum_{j=1}^{n} < f,\phi_j > \phi_j, f - \sum_{j=1}^{n} < f,\phi_j > \phi_j > \geq 0.$$

After expanding the inequality (1.12) becomes

$$(1.13) \qquad \sum_{j=1}^{n} | < f,\phi_j > |^2 \leq \| f \|^2$$

The right side of (1.13) is independent of $n$ and hence must hold in the limit as $n \to \infty$. The result is the Bessel inequality

$$(1.14) \qquad \sum_{j=1}^{\infty} | < f,\phi_j > |^2 \leq \| f \|^2 .$$

Since the Bessel inequality implies the convergence of the infinite series, the individual terms must go to zero as $j \to \infty$ Q.E.D.

The Bessel inequality (1.14) holds for every infinite orthonormal sequence. However, in the particular case in which equality holds for every $f \epsilon H$, the set $\{\phi_n\}$ is said to be a complete orthonormal set, i.e.,

Definition: If

$$(1.15) \qquad \sum_{j=1}^{\infty} | < f,\phi_j > |^2 = \| f \|^2$$

for every $f \epsilon H$ then $\{\phi_n\}$ is a complete orthonormal set.

These complete orthonormal sets play an important role in Hilbert space theory since it can be shown that every vector in the space can be expanded in terms of these vectors. In particular if $f \epsilon H$ then

$$(1.16) \qquad \lim_{n \to \infty} \| f - \sum_{j=1}^{n} < f,\phi_j > \phi_j \| = 0$$

so that

$$(1.17) \qquad f = \sum_{j=1}^{\infty} < f,\phi_j > \phi_j .$$

There are many examples of complete orthonormal sets for $\mathcal{L}_2(a,b)$. Thus for the real Hilbert space $\mathcal{L}_2(0,\pi)$ the set $\{\sqrt{\frac{2}{\pi}}\sin nx\}$ is a complete orthonormal set.

Definition: S is a closed linear manifold in H if

1) S is a subspace of H

2) S contains its limit points

The following theorem is stated without proof.

Theorem (1.2) (Projection Theorem): Let M be a closed linear manifold in H and let $M^{\perp}$ be the orthogonal complement of M. Every vector $x \epsilon H$ has a unique representation $x = x_p + z$ when $x_p \epsilon M$ and $z \epsilon M^{\perp}$.

<u>Linear Functionals</u>

The preceding remarks have been devoted to describing various properties of Banach and Hilbert spaces. It is now of interest to describe the properties of various transformations defined on these spaces.

Definition: T(x) is a functional on H if T(x) assigns to every element $x \epsilon H$ a complex number.

Definition: T(x) is a linear functional on H if $T(x_1 + x_2) = T(x_1) + T(x_2)$ and $T(\alpha x) = \alpha T(x)$.

Definition: T(x) is bounded on H if there exists a real number C such that $|T(x)| \leq C \|x\|$ for all $x \epsilon H$.

The minimum value of C for which $|T(x)| \leq C \|x\|$ for every $x \epsilon H$ is called the norm of T and written $\|T\|$. Thus

$$(1.18) \qquad \|T\| = \operatorname*{lub}_{\|x\| \neq 0} \frac{|T(x)|}{\|x\|} .$$

Definition: $T$ is continuous at $x$ if $x_n \to x$ implies that $T(x_n) \to T(x)$.

In order to prove the continuity of a linear functional in $H$, it suffices to prove the continuity at $x = 0$. This is simply a consequence of the fact that if $x_n \to x$ then $T(x_n - x) \to 0$ if $T$ is continuous at the origin; the result follows from the linearity of $T$. In the above definitions we have distinguished between the boundedness and continuity of a functional. However, for linear functionals this distinction is unnecessary.

Theorem (1.3): If $T$ is linear then $T$ is bounded iff $T$ is continuous.

Proof: If $T$ is bounded and $x_n \to x$, it follows that $|T(x_n) - T(x)| = |T(x_n - x)| \leq \|T\| \; \|x_n - x\| \to 0$. Thus $T$ is continuous. Conversely, if $T$ is continuous but $T$ is not bounded, it follows that there exists a sequence $x_n$ such that $|T(x_n)| \geq n \|x_n\|$. However, $y_n = x_n/(n \|x_n\|) \to 0$ as $n \to \infty$, so that, in view of the continuity of $T$, $T(y_n) \to 0$. On the other hand, $|T(y_n)| = |T(x_n)|/(n \|x_n\|) \geq n \|x_n\|/(n \|x_n\|) = 1$.
This contradiction proves the theorem Q.E.D.

The most commonly cited example of a continuous linear functional is the inner product $T(x) = \langle x,y \rangle$ for a fixed $y \epsilon H$. However, the inner product is more than an example of a continuous linear functional in that <u>every</u> continuous linear functional can be written as an inner product.

Theorem (1.4) (Riesz Representation Theorem): Every continuous linear functional $T(x)$ on $H$ has a representation $T(x) = \langle x,g \rangle$, where $g \epsilon H$ is uniquely determined by $T$.

Proof: Let $N = \{x \epsilon H \mid T|x| = 0\}$, i.e., $N$ is the null space of $T$.

It is easily verified that $N$ is a closed linear manifold. If $N = H$ then $T(x) = \langle x,0 \rangle$, so that it can be assumed that the orthogonal complement of $N$, say $N^{\perp}$, is not empty. Let $f_o \epsilon N^{\perp}$ such that $\|f_o\| = 1$. Since

$T(xT(f_0) - f_0 T(x)) = T(f_0)T(x) - T(f_0)T(x) = 0$ it follows that $xT(f_0) - f_0 T(x) \epsilon N^{\perp}$ or equivalently $<xT(f_0) - f_0 T(x), f_0> = 0$. Thus $T(x) = T(f_0) <x,f_0> = <x,\overline{T}(f_0)f_0>$. Therefore it suffices to choose $g = \overline{T}(f_0)f_0$. In order to prove the uniqueness of $g$ simply note that if $<x,f> = <x,g>$ for all $x \epsilon H$, then $<x,g - f> = 0$ when $x$ is chosen equal to $g - f$, i.e., $\|g - f\| = 0$ or $g = f$ Q.E.D.

## Linear Transformations

The purpose of this section is to discuss various mappings defined on a Hilbert space $H$ which are more general than linear functionals.

Definition: $A$ is a transformation on $H$ if $A$ assigns to each $x \epsilon H$ another element $y = Ax \epsilon H$.

Definition: $A$ is linear if $A(x_1 + x_2) = Ax_1 + Ax_2$ and $A(\alpha x) = \alpha Ax$.

Definition: A linear transformation $A$ is bounded if there exists a real number $C$ such that $\| Ax \| \leq C \| x \|$ for all $x \epsilon H$.

The minimum $C$ for which $\| Ax \| \leq C \| x \|$ is called the norm of $A$, written $\| A \|$. Thus

$$\| A \| = \operatorname*{lub}_{\| x \| \neq 0} \frac{\| Ax \|}{\| x \|} = \operatorname*{lub}_{\| x \| \neq 0} \frac{<Ax,Ax>^{\frac{1}{2}}}{<x,x>^{\frac{1}{2}}} .$$

Definition: $A$ is continuous at $x$ if $x_n \to x$ implies $Ax_n \to Ax$.

Just as for continuous linear functionals, it is easily shown that (i) if $A$ is continuous at $x = 0$ then $A$ is continuous in $H$, and (ii) if $A$ is linear then continuity and boundedness are equivalent.

If $A$ is a bounded linear operator, it follows that for each $y \epsilon H$ the functional $<Ax,y>$ is bounded and linear. Thus the Riesz representation theorem guarantees the existence of a unique element $g \epsilon H$ such that $< Ax, y > = < x, g >$. The vector $g$ clearly depends on $y$ This dependence can be expressed as $g = A^* y$. The transformation $A^*$ is defined for all $y \epsilon H$ and is linear

Definition: $A^*$ is the adjoint of $A$.

Theorem (1.5): $\| A \| = \| A^* \|$.

Proof: For all $x, y \epsilon H$ the operators $A$ and $A^*$ are related by $<Ax, y> = <x, A^* y>$. In particular, if $x = A^* y$ this identity implies that $\| A^* y \|^2 = <AA^* y, y> \leq \| AA^* y \| \| y \| \leq \| A \| \| A^* y \| \| y \|$, i.e., $\| A^* y \| \leq \| A \| \| y \|$. Thus $\| A^* \| \leq \| A \|$. A repetition of this argument with $x = Ay$ yields $\| A \| \leq \| A^* \|$. Q.E.D.

Definition: If $A = A^*$ then $A$ is symmetric.

Definition: A set $S$ is bounded if there exists a real number $C$ such that $\| x \| \leq C$ for all $x \epsilon S$.

Definition: A set $S$ is compact if every infinite sequence $\{x_n\}$ in $S$ contains a convergent subsequence.

Definition: Let $A$ be a linear transformation defined on a complete bounded set $S$. If $R_A = \{Ax / x \epsilon S\}$ is compact, then $A$ is completely continuous.

Completely continuous operators are a proper subset of the bounded operators. In fact it is easily shown that every completely continuous operator is bounded, but there are bounded operators, e.g., the identity operator, which are not completely continuous. Even so, there are a wide variety of operators occurring in the applications which are completely continuous.

Definition: A non-zero vector $x$ is an eigenvector of $A$ if there exists a (complex) number $\lambda$, called the eigenvalue, such that $Ax - \lambda x = 0$.

The object in what follows is to prove the existence of eigenvalues and eigenvectors for symmetric, completely continuous operators.

Theorem (1.6): The eigenvalues (if they exist) of a symmetric, bounded, linear transformation are real.

Proof: Let $x$ be an eigenvector and $\lambda$ the corresponding eigenvalue for $A$. It follows that $<Ax,x> = \lambda <x,x>$. Moreover, the symmetry of $A$ implies that $<Ax,x> = <x,A^*x> = <x,Ax> = \overline{<Ax,x>} = \overline{\lambda}<x,x>$. Thus $\lambda = \overline{\lambda}$, or equivalently $\lambda$ is real. Q.E.D.

Theorem (1.7): If $A$ is bounded and $\|x\| \neq 0$

$$(1.19) \qquad \frac{|<Ax,x>|}{\|x\|^2} \leq \|A\|.$$

If $\|x\| = 1$

$$(1.20) \qquad |<Ax,x>| \leq \|A\|.$$

Proof: The proof is an immediate consequence of the Schwartz Inequality, i.e.,

$$(1.21) \qquad |<Ax,x>| \leq \|Ax\| \; \|x\| \leq \|A\| \|x\|^2.$$

Corollary (1.1): Define

$$(1.22) \qquad M_A = \underset{\|x\| \neq 0}{\text{lub}} \; \frac{|<Ax,x>|}{\|x\|^2}.$$

Then

$$(1.23) \qquad M_A \leq \|A\|.$$

Theorem (1.8): If $\lambda$ is an eigenvalue of $A$, then $|\lambda| \leq \|A\|$.

Proof: $\dfrac{|<Ax,x>|}{\|x\|^2} = |\lambda| \dfrac{<x,x>}{\|x\|^2} = |\lambda|$.

Theorem (1.9): If $A$ is symmetric, then $M_A = \|A\|$.

Proof: In view of corollary (1.1), it suffices to show that $\| A \| \leq M_A$.

This fact follows from the identity

$$(1.24) \qquad \langle A(x+y), (x+y) \rangle - \langle A(x-y), x-y \rangle = 2\langle Ax, y \rangle + 2\langle Ay, x \rangle$$

where it should be noted that $\langle A(x+y), (x+y) \rangle$ and $\langle A(x-y), (x-y) \rangle$ are real numbers.

Since (cf. Theorem (1.7))

$$(1.25) \qquad \langle A(x+y), (x+y) \rangle \leq M_A \| x+y \|^2$$

and

$$(1.26) \qquad \langle A(x-y), (x-y) \rangle \geq - M_A \| x-y \|^2$$

the identity (1.24) can be replaced by the estimate

$$(1.27) \qquad \langle Ax, y \rangle + \langle Ay, x \rangle \leq M_A ( \| x \|^2 + \| y \|^2).$$

If $y$ is chosen equal to $Ax \| x \| / \| Ax \|$ and the symmetry of $A$ is used, the theorem follows. Q.E.D.

Theorem (1.10): If $A$ is symmetric, there exists a sequence $\{x_k\}$ with $\| x_k \| = 1$ such that

$$(1.28) \qquad \lim_{k \to \infty} (Ax_k - \lambda_1 x_k) = 0$$

where $\lambda_1$ is either $+ \| A \|$ or $- \| A \|$.

Proof: Theorem (1.9) implies the existence of a sequence $\{z_k\}$ such that $\| z_k \| = 1$ and

$$(1.29) \qquad \lim_{k \to \infty} | \langle Az_k, z_k \rangle | = \| A \|.$$

Since $\langle Az_k, z_k \rangle$ is real, a subsequence $\{x_k\}$ can be chosen such that $\langle Ax_k, x_k \rangle$ converges to either $\pm \| A \|$. Call this limit $\lambda_1$, i.e.,

(1.30) $$\lim_{k\to\infty} \langle Ax_k, x_k \rangle = \lambda_1 .$$

This result can be used to estimate $\| Ax_k - \lambda_1 x_k \|^2$. Indeed (1.30) implies that

(1.31) $$\lim_{k\to\infty} \| Ax_k - \lambda_1 x_k \| = 0 . \quad \text{Q.E.D.}$$

Theorem (1.10) by no means implies the existence of an eigenvector since it is not known that the sequence $\{x_k\}$ converges. In fact, to prove the existence of an eigenvector, and hence an eigenvalue, it will be necessary to require stronger conditions on $A$ than simple boundedness.

Theorem (1.11): If $A$ is symmetric and completely continuous, at least one of the numbers $+\|A\|$ or $-\|A\|$ is an eigenvalue. Moreover, no eigenvalue has larger absolute value.

Proof: Since $A$ is completely continuous, there exists a subsequence of $\{x_k\}$, say $\{u_k\}$, such that $Au_k$ converges. Since $Au_k - \lambda_1 u_k \to 0$ it follows that $\{u_k\}$ converges to some vector $u$. It is an immediate consequence that $Au - \lambda u = 0$. Q.E.D.

Theorem (1.11) shows that solutions of the equation $Au - \lambda u = 0$ have at least one bifurcation point. Moreover, this bifurcation point occurs for $|\lambda_1|$ less than or equal to $\|A\|$. In view of the above results, it is important to have an elementary criterion for determining when an operator is completely continuous. This is the purpose of the next theorem.

Theorem (1.12): If $A$ can be approximated in norm by completely continuous operators, then $A$ is completely continuous.

Proof: Assume there exists a sequence of completely continuous operators $\{A_n\}$ such that $\|A - A_n\| \to 0$ as $n \to \infty$. Let $\{x_k\}$ be any bounded sequence. Since $A_1$ is completely continuous, there exists a subsequence $\{x_k^{(1)}\}$ such that $A_1 x_k^{(1)}$ converges. Similarly, since $A_2$ is completely continuous, there exists a subsequence of $\{x_k^{(1)}\}$, say $\{x_k^{(2)}\}$ such that $A_2 x_k^{(2)}$ converges. This procedure can be repeated for each operator $A_n$. The diagonal sequence

$\{x_k^{(k)}\}$ has the property that $A_n x_k^{(k)}$ converges for each $n$. In order to show that $A$ is itself completely continuous, it suffices to show that $Ax_k^{(k)}$ converges. This is a consequence of the fact that

$$(1.32) \quad \| Ax_n^{(n)} - Ax_m^{(m)} \| \leq \| Ax_n^{(n)} - A_k x_n^{(n)} \|$$

$$+ \| A_k x_n^{(n)} - A_k x_m^{(m)} \| + \| A_k x_m^{(m)} - Ax_m^{(m)} \|$$

so that

$$(1.33) \quad \| Ax_n^{(n)} - Ax_m^{(m)} \| \leq \| A - A_k \| ( \| x_n^{(n)} \|$$

$$+ \| x_m^{(m)} \| ) + \| A_k x_n^{(n)} - A_k x_m^{(m)} \| .$$

The right side of (1.32) can be made arbitrarily small by first choosing $k$ and then choosing $m$ and $n$. Thus $\{ Ax_n^{(n)} \}$ is a Cauchy sequence and hence $Ax_n^{(n)}$ converges. Q.E.D.

## Linear Integral Equations

The object of this section is to discuss the existence of eigenvalues and eigenvectors for linear integral equations of the form

$$(1.34) \quad Ku - \mu u = \int_a^b k(x,\xi)u(\xi)d\xi - \mu u(x) = 0$$

Definition: If

$$(1.35) \quad \int_a^b \int_a^b |k(x,\xi)|^2 dx d\xi < \infty$$

then $K$ is a Hilbert-Schmidt integral operator.

It will be shown below that if $K$ is Hilbert-Schmidt, then $K$ is completely continuous.

Definition:

$$(1.36) \quad su = \int_a^b A(x,\xi)u(\xi)d\xi = \int_a^b \sum_{i=1}^{n} p_i(x)q_i(\xi)u(\xi)d\xi$$

is a degenerate operator if

$$(1.37) \qquad \| p_i \|^2 = \int_a^b |p_i(x)|^2 dx < \infty, \quad \| q_i \|^2 = \int_a^b |q_i(\xi)|^2 d\xi < \infty$$

$$i = 1,2,\ldots,n.$$

Theorem (1.13):

$$(1.38) \qquad su = \sum_{i=1}^{n} p_i(x) \int_a^b q_i(\xi) u(\xi) d\xi$$

is completely continuous.

Proof: $s$ is bounded since

$$(1.39) \qquad \| su \| \leq \left( \sum_{i=1}^{n} \| p_i \| \, \| q_i \| \right) \| u \| .$$

In addition, the range of $s$ is finite dimensional. However, every complete, bounded, finite dimensional subspace is compact. Q.E.D.

In order to show that $K$ is completely continuous, it only remains to show that $K$ can be approximate by degenerate kernels. This is an immediate consequence of the following lemma which is stated without proof.

Lemma (1.2): If $\{\phi_i(x)\}$ is a complete orthonormal set on $a \leq x \leq b$ and $\{\psi_j(\xi)\}$ is a complete orthonormal set on $c \leq \xi \leq d$, then $\{\phi_i(x)\psi_j(\xi)\}$ is a complete orthonormal set on the rectangle $a \leq x \leq b$, $c \leq \xi \leq d$.

Theorem (1.14): A Hilbert-Schmidt integral operator is completely continuous.

Proof: Assume $\{\phi_i(x)\}$ is a complete orthonormal set on $a \leq x \leq b$. It is a consequence of lemma (1.2) that $\{\phi_i(x)\overline{\phi}_j(\xi)\}$ is a complete orthonormal set on the square $a \leq x, \xi \leq b$, i.e., for every function $k(x,\xi)$ defined on the square $a \leq x, \xi \leq b$ and satisfying the condition (1.35), it is true that

$$(1.40) \qquad \lim_{n\to\infty} \| k(x,\xi) - \sum_{i,j=1}^{n} a_{ij} \phi_i(x) \overline{\phi}_j(\xi) \| = 0,$$

$$a_{ij} = \int_a^b \int_a^b k(x,\xi) \overline{\phi}_j(\xi) dx d\xi .$$

Thus degenerate kernels $s_N$ can be chosen such that

$$\| K - s_n \| \to 0. \tag{1.41}$$

Theorem (1.12) implies the complete continuity of K. Q.E.D

It is easily shown that a sufficient condition for K to be symmetric is that $k(x,\xi) = \overline{k}(\xi,x)$. If K is symmetric and Hilbert-Schmidt, it follows from the development of the preceding section that K has at least one real eigenvalue and eigenfunction. This result can be stated explicitly as

Theorem (1.15): If K is symmetric, Hilbert-Schmidt operator, the extremal problem

$$\max_{\| u \| = 1} | < Ku, u > | \tag{1.42}$$

has a solution. Any solution is a normalized eigenfunction corresponding to an eigenvalue $\mu_1$. Furthermore,

$$\mu_1 = \max_{\| u \| = 1} | < Ku, u > |. \tag{1.43}$$

Let $\phi_1$ be the normalized eigenfunction corresponding to the eigenvalue $\mu_1$ described in theorem (1.15). Define the operator $K_1$ corresponding to the kernel

$$k_1(x,\xi) = k(x,\xi) - \mu_1 \phi_1(x)\overline{\phi}_1(\xi). \tag{1.44}$$

The operator $K_1$ is clearly symmetric and Hilbert-Schmidt. Moreover, it is easily verified that $K_1$ has the following properties:

(1) If $<u, \phi_1> = 0$ then $K_1 u = Ku$,

(2) If $u = c\phi_1$ then $K_1 u = 0$,

(3) $<K_1 u, \phi_1> = 0$, and

(4) If $K_1 u = \lambda u (\lambda \neq 0)$ then $Ku = \lambda u$.

Theorem (1.11) can now be applied to the operator $K_2$.

Theorem (1.16): The extremal problem

$$(1.45) \qquad \max_{\|u\|=1} |\langle K_1 u, u \rangle|$$

has a solution. Any solution is a normalized eigenfunction of $K_1$ corresponding to an eigenvalue $\mu_2$ where

$$(1.46) \qquad \mu_2 = \max_{\|u\|=1} |\langle K_1 u, u \rangle|.$$

Moreover, the eigenfunction $\phi_2$ satisfies $\langle \phi_1 \phi_2 \rangle = 0$. Since the solution of (1.45) satisfies the condition $\langle \phi_1 \phi_2 \rangle = 0$ it suffices to maximize (1.45) under the condition $\langle u, \phi_1 \rangle = 0$. However, for all functions $u$ such that $\langle u, \phi_1 \rangle = 0$ property (1) above implies $K_1 u = Ku$. Thus theorem (1.16) can be restated as

Theorem (1.17): The extremal problem

$$(1.47) \qquad \max_{\substack{\|u\|=1 \\ \langle u, \phi_1 \rangle = 0}} |\langle Ku, u \rangle|$$

has a solution. The solution is a normalized eigenfunction $\phi_2$ corresponding to an eigenvalue $\mu_2$, where

$$\mu_2 \leq \mu_1 \quad \text{and} \quad \langle \phi_2, \phi_1 \rangle = 0.$$

This procedure can be continued so that at the n'th stop $K_n$ is defined by the kernel

$$(1.48) \qquad k_n(x,\xi) = k(x,\xi) - \sum_{i=1}^{n} \mu_i \phi_i(x) \overline{\phi}_i(\xi).$$

Theorem (1.18): The extremal problem

$$(1.49) \qquad \max_{\substack{\|u\|=1 \\ \langle u_i \phi_i \rangle = 0 \quad i = 1,\dots,n}} |\langle Ku, u \rangle|$$

has a solution. The solution is a normalized eigenfunction $\phi_{n+1}$ corresponding to an eigenvalue $\mu_{n+1}$, where

$$\mu_{n+1} \leq \mu_n \leq \cdots \leq \mu_1 \quad \text{and} \quad <\phi_{n+1},\phi_1> = 0, \quad i = 1,\ldots,n.$$

Actually at each stop in the above described procedure, it has been implicitly assumed that $K_n \not\equiv 0$. However, this can only happen if $K$ is degenerate. Thus if $K$ is nondegenerate, the equation has infinitely many eigenvalues and corresponding independent eigenfunctions.

Theorem (1.19): The number of independent eigenfunctions corresponding to a non-zero eigenvalue is finite.

Proof: Assume $\psi_1,\ldots,\psi_n$ are eigenfunctions corresponding to $\mu$. Every linear combination

$$\sum_{i=1}^{n} c_i \psi_i$$

is also an eigenfunction. Therefore, the set of eigenvectors corresponding to $\mu$ is a linear space. Let $\phi_1,\ldots,\phi_n$ be an orthonormal basis for this space. It is a consequence of

$$(1.50) \qquad \int_a^b \int_a^b |k(x,\xi)| - \sum_{i=1}^{n} \mu\phi_i(x)\overline{\phi}_i(\xi)|^2 dx d\xi \geq 0$$

that

$$(1.51) \qquad n\mu^2 \leq \int_a^b \int_a^b |k(x,\xi)^2 dx d\xi$$

since $K$ is a Hilbert-Schmidt kernel and $\mu$ is not zero, the inequality (1.51) implies that $n$ is finite. Q.E.D.

Theorem (1.20): If there are infinitely many non-zero eigenvalues, they converge only at the origin.

Proof: An argument similar to that employed in proving theorem (1.19) shows that

$$(1.52)\qquad \sum_{i=1}^{\infty} \mu_i^{\,2} \le \int_a^b \int_a^b |k(x,\xi)|^2\, dx d\xi$$

thus $\mu_n \to 0$. Q.E.D.

In summary, it has been shown that the integral equation (1.34) with $K$ symmetric and Hilbert-Schmidt (nondegenerate) has infinitely many bifurcation points. Moreover, the bifurcation points converge to zero and no bifurcation point, with the possible exception of zero, corresponds to more than a finite number of eigenfunctions. If $K$ is degenerate, then there are a finite number of bifurcation points and zero is a bifurcation point corresponding to infinitely many eigenfunctions.

Linear Differential Equations

The results obtained in the preceding section for linear integral equations are readily applicable to linear differential equations. Consider the differential equation

$$(1.53)\qquad Lu + \lambda u = 0 \quad a < x < b$$

where

$$(1.54)\qquad L = a_0(x)\frac{d^2}{dx^2} + a_1(x)\frac{d}{dx} + a_2(x)$$

with boundary data

$$(1.55)\qquad B_1(u) = \alpha_1 u(a) + \alpha_2 u'(a) = 0$$

$$B_2(u) = \beta_1 u(b) + \beta_2 u'(b) = 0$$

It is assumed that $a_0$, $a_1$ and $a_2$ are continuous and that $a_0$ does not vanish in the interval $a \le x \le b$. Under these conditions a well known theorem guarantees the existence and uniqueness of solutions to the initial value problem for $Lw = 0$. In the following remarks it will be assumed that the problem $Lw = 0$ with boundary data (1.55) has only the trivial solution.

Definition: $g(x,\xi)$ is a Green's function for $L$ if

(1.56) $$Lg(x,\xi) = 0, \quad a \le x < \xi, \quad \xi < x \le b,$$

(1.57) $$B_1(g(x,\xi)) = B_2(g(x,\xi)) = 0$$

$g(x,\xi)$ is continuous at $x = \xi$, and

(1.58) $$\frac{dg}{dx}\Big|_{x=\xi_+} - \frac{d\xi}{dx}\Big|_{x=\xi_-} = \frac{1}{a_0(\xi)} .$$

It may be verified by differentiation that if the Green's function exists, the integral equation

(1.59) $$u = \lambda \int_a^b g(x,\xi)u(\xi)d\xi = \lambda Gu$$

is equivalent to equ. (1.53) with boundary data (1.57). However, the existence of the Green's function follows from the existence and uniqueness of solutions to the initial value problem. For example, if $u_1$ is a solution of (1.53) satisfying $u_2(b) = -\beta_2, u_2'(b) = \beta_1$, then the Green's function is given by

(1.60) $$g(x,\xi) = \begin{cases} \dfrac{u_1(x)u_2(\xi)}{a_0(\xi)W(u_1,u_2,\xi)} & a \le x \le \xi \\[2ex] \dfrac{u_1(\xi)u_2(x)}{a_0(\xi)W(u_1,u_2,\xi)} & \xi \le x \le b \end{cases}$$

where

(1.61) $$W(u_1,u_2,\xi) = u_1(\xi)u_2'(\xi) - u_1'(\xi)u_2(\xi)$$

does not vanish since $u_1$ and $u_2$ are linearly independent. This linear independence is a consequence of the assumption that the only solution of $Lu = 0$ with $B_1(u) = B_2(u) = 0$ is $u \equiv 0$. The Green's function is symmetric and continuous. Thus the integral equation (1.59) is a symmetric Hilbert-Schmidt integral equation and, in fact, can be rewritten

(1.62) $$\mu u = Gu$$

with $\mu = 1/\lambda$ since by assumption $\lambda = 0$ is not an eigenvalue.

Theorem (1.21): $\mu = 0$ ($\lambda = \pm \infty$) is not an eigenvalue.

Proof: If $\mu = 0$ is an eigenvalue, there must exist a function f such that

(1.63) $$Gf = 0.$$

However, the solution of the equation $Lu = f$ is given by $u = Gf = 0$. This implies $f = 0$. Q.E.D.

Since $\mu = 0$ is not an eigenvalue, equ. (1.62) has infinitely many non-zero eigenvalues converging to zero. Equivalently, the differential equ. (1.53) with boundary data (1.54) has infinitely many bifurcation points $\lambda_n$ converging to $\pm \infty$. Moreover, it follows from our results in the preceding section that none of these bifurcation points correspond to more than a finite number of solutions.

Sturmian Theory

Up to this point in these lecture notes, the existence of branches of solutions to various linear equations have been discussed while saying little about the qualitative behaviour of these solutions. However, it is well known that solutions of the equation

(1.64) $$Lw = \frac{d}{dx}p(x)\frac{dw}{dx} - q(x)w = 0$$

are twice continuously differentiable on any interval $a \leq x \leq b$ which $p'(x)$ and $q(x)$ are continuous and p does not vanish. In certain cases it is possible, by comparing equ. (1.64) to equations with known solutions, to obtain estimates on the number of zeros of solutions to equ. (1.64) in the interval $a < x < b$.

In order to guarantee the existence and uniqueness of solutions to equ. (1.64) on the interval $a \leq x \leq b$, it will be assumed that p does not vanish on this interval. Thus there is no loss of generality in assuming $p > 0$. However, results to be obtained below apply equally well to the case in which p vanishes at either end of the interval if the existence of a continuous solution can be guaranteed.

Theorem (1.22): In any finite interval $a \leq x \leq b$ non-trivial solutions u of $Lu = 0$ have at most a finite number of zero's.

Proof: If $u$ has an infinite number of zeros on the interval $a \leq x \leq b$ then there must be a limit point of zeros in that interval. It follows that both $u$ and $u'$ vanish there. The only solution of $Lu = 0$ with both $u$ and $u'$ zero at a point is $u = 0$. This contradiction proves the theorem. Q.E.D.

Theorem (1.23): Let $u_1$ and $u_2$ be linearly independent solutions of $Lu = 0$. The zeros of $u_1$ and $u_2$ separate each other.

Proof: Assume $x_1$ and $x_2$ are consecutive zeros of $u_1$. The function of $u_2$ cannot be zero at either $x_1$ or $x_2$ since it would follow that the Wronskian would vanish, and hence $u_1$ and $u_2$ would be independent. If $u_2$ does not vanish in the interior of the interval $a \leq x \leq b$, then the function $v = u_1/u_2$ is continuous and differentiable on this interval. However, since $u_1$ vanishes at both $x_1$ and $x_2$, Rolle's theorem implies that $u' = 0$ for some $x$ in the interval $a < x < b$, i.e.,

$$(1.65) \qquad v'(x) = \frac{u_1'(x)u_2(x) - u_2'(x)u_1(x)}{u_2(x)^2} = 0.$$

On the other hand, $v'(x)$ cannot vanish on the interval since the numerator of equ. (1.65) is simply the Wronskian. Thus $u_2$ must have a zero between $x_1$ and $x_2$. Q E.D.

It may be noted that $u_2$ cannot have two zeros in the interval since theorem (1.23) also applies to $u_2$. Thus the general situation must be as depicted in fig. 1.6.

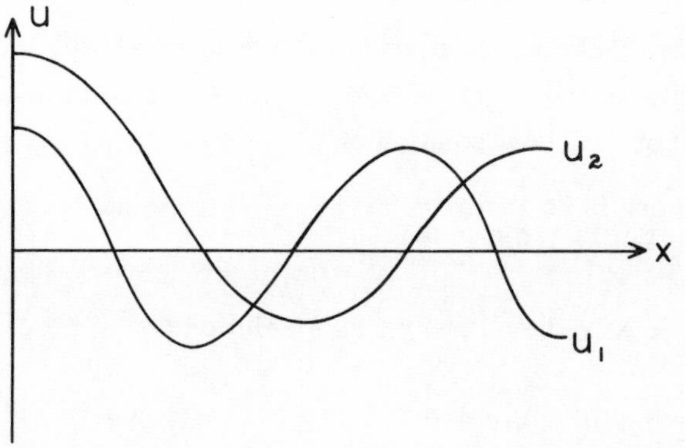


fig. 1.6

Theorem (1.23) compares the zeros of distinct solutions of equ. (1.64). The object now is to compare zeros of solutions of different equations.

Theorem (1.24): Let $u$ be a solution of

$$(1.66) \qquad L_1u = (p(x)u')' - q_1(x)u = 0$$

on the interval $a \leq x \leq b$. Let $v$ be a solution of

$$(1.67) \qquad L_2v = (p(x)v')' - q_2v = 0.$$

Assume $q_1 \geq q_2$ on this interval and that $q_1 \not\equiv q_2$. If $x_1$ and $x_2$ are consecutive zeros of $u$, there exists an $x$ such that $a < x < b$ and $v(x) = 0$.

Proof: Assume $v$ has no zeros in the interval $a < x < b$. It can then be assumed that $u > 0$ and $v > 0$ in the interior of the interval. The identity

$$(1.68) \qquad vL_1u - uL_2v = \frac{d}{dx}p(u'v - v'u) - (q_1 - q_2)uv = 0$$

implies

$$(1.69) \qquad p(x)u'(x)v(x)\Big|_{x_1}^{x_2} = \int_{x_1}^{x_2}(q_1 - q_2)uvdx.$$

The right side of equ. (1.69) is definitely positive. Since $u'(x_2)$ is negative and $u'(x_1)$ is positive (cf. fig. 1.7), the left side of equ. (1.69) is negative. This contradiction implies that $v$ must vanish in the interval. Q.E.D.

Theorem (1.25): Let $u$ be a solution of

$$(1.70) \qquad L_1u = (p_1u')' - q_1u = 0$$

on the interval $a \leq x \leq b$. Let $v$ be a solution of

$$(1.71) \qquad L_2v = (p_2v')' - q_2v = 0.$$

Assume $p_1 > p_2$ and $q_1 \geq q_2$ on the interval (it is assumed that equ. (1.70) and equ. (1.71) are not identical on $a \leq x \leq b$). If $x_1$ and $x_2$ are consecutive zeros of $u$, there exists an $x$ such that $a < x < b$ and $v(x) = 0$.

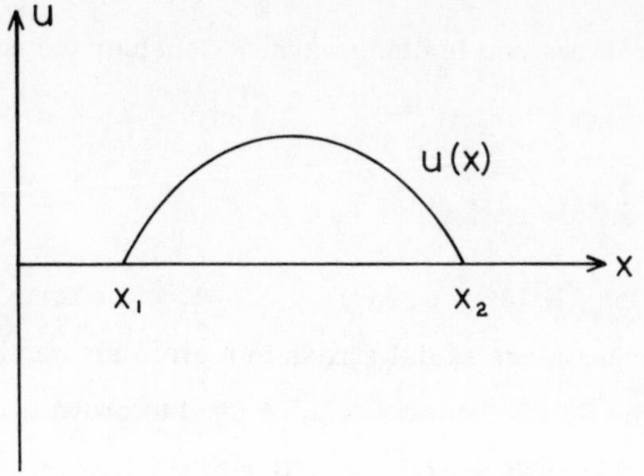


fig. 1.7

Proof: The result follows from the identity

$$(1.72) \qquad \frac{d}{dx}\frac{u}{v}(p_1u'v - p_2uv') = (q_1 - q_2)u^2 + (p_1 - p_2)u'^2 + p_2\frac{(u'v - uv')^2}{v^2} .$$

Assume $v(x_1)$ and $v(x_2)$ are not zero. The identity (1.72) implies

$$(1.73) \qquad 0 = \int_{x_1}^{x_2}(q_1 - q_2)u^2dx + \int_{x_1}^{x_2}(p_1 - p_2)u'^2dx + \int_{x_1}^{x_2}p_2\frac{(u'v - uv')^2}{v^2}dx.$$

The right side of (1.73) is positive. This is a contradiction unless either or both $v(x_1)$ and $v(x_2)$ are zero. However, even in this case L'Hospital's rule implies that if $v(x_1) = 0$

$$\lim_{x \to x_1}\frac{u}{v}(p_1u'v - p_2uv') = 0$$

Q.E.D.

The preceding two (Sturm) theorems can be used to obtain information on the oscillatory properties of solutions to linear differential equations with non-constant coefficients. Thus it is a trivial exercise to show that if $q > 0$, then solutions of equ. (1.64) are non-oscillatory. In addition, if $P_M \geq p \geq P_m > 0$ and $q_M \geq q > 0$, the Sturm theorems can be used to estimate the number of zeros of eigenfunctions of

$$(1.74) \qquad (pu')' + (\lambda + q(x_1)u) = 0.$$

The Sturm theorems can be used in other ways. Consider the nonlinear differential equation

$$(1.75) \qquad \alpha^2(\alpha'' + \frac{3}{x}\alpha') + 2 = 0$$

with boundary conditions $\alpha'(0) = 0$, $\alpha(1) = \lambda > 0$. The term $\alpha$ in equ. (1.75) is essentially the dimensionless radial stress in a circular membrane deformed by normal pressure. The Sturm theorems can be used to prove that this problem has a solution if $\lambda$ is sufficiently large. In order to see how this is done, rewrite equ. (1.75) as an integral equation

$$(1.76) \qquad \alpha = \lambda + \int_0^x \frac{\tau^3/x^2 - \tau^3}{\alpha^2}\, d\tau + \int_x^1 \frac{\tau - \tau^3}{\alpha^2}\, d\tau = \lambda + G\,\frac{1}{\alpha^2}\,.$$

Define an iteration scheme $\alpha_{n+1} = \lambda + G(1/\alpha_n^2)$ with $\alpha_0 = \lambda$. It is easily seen that the iterates have the property that

$$(1.77) \qquad \lambda = \alpha_0 \leq \alpha_2 \leq \cdots \leq \alpha_{2n} \leq \alpha_{2n+1} \leq \cdots \leq \alpha_3 \leq \alpha_1$$
$$= \lambda + \frac{1}{4\lambda^2}(1 - x^2).$$

Since the derivatives of the iterates are also bounded independent of $n$, $\alpha_{2n}$ must converge to a function $\alpha_-$ and $\alpha_{2n+1}$ must converge to a function $\alpha_+$, where $\alpha_+ \geq \alpha_-$. The functions $\alpha_+$ and $\alpha_-$ are solutions of $\alpha_+ = \lambda + G(1/\alpha_-^2)$ and $\alpha_- = \lambda + G(1/\alpha_+^2)$. Thus the existence of a solution to equ. (1.75) satisfying the prescribed boundary conditions will follow if it can be shown that $\alpha_+ = \alpha_-$. The difference function $v = \alpha_+ - \alpha_-$ satisfies the differential equation

$$(1.78) \qquad v'' + \frac{3}{x}v' + 2\,\frac{\alpha_+ + \alpha_-}{\alpha_+^2\alpha_-^2}\,v = 0$$

and the homogeneous boundary conditions $v'(0) = 0$ and $v(1) = 0$. However, since

$$\frac{\alpha_+ + \alpha_-}{\alpha_+^2 \alpha_-^2} \leq \frac{2}{\lambda^3}$$

theorem (1.24) implies that solutions of the equation

$$(1.79) \qquad w'' + \frac{3}{x}w' + \frac{4}{\lambda^3}w = 0$$

must have a zero between the origin and the first zero of any solution to equ. (1.78). A solution of equ. (1.79) is

$$(1.80) \qquad w = \frac{J_1\left(\frac{2}{\lambda^{3/2}}x\right)}{x}$$

where $J_1$ is the Bessel function of first order. Thus if $2/\lambda^{3/2} < j_{11}$ where $j_{11}$ is the smallest zero of $J_1$, equ. (1.80) has no zeros in the interval $0 \leq x \leq 1$, and hence no non-trivial solution of equ. (1.79) can have a zero in this interval, i.e., the only solution of equ. (1.79) satisfying the homogeneous boundary conditions is $v \equiv 0$. We conclude that equ. (1.75) has a solution if $\lambda^{3/2} > 2/j_{11}$.

References

1 E.L. Ince, Ordinary Differential Equations, Dover Publications, New York, 1956.

2 I. Stakgold, Boundary Value Problems of Mathematical Physics, Vol.I, MacMillan, New York, 1967.

Other References

3 W.J Akhiezer and I.M. Glazman, Theory of Linear Operators in Hilbert Spaces, Ungar, New York, 1955.

4 E.A. Coddington and N. Levinson, Theory of Ordinary Differential Equations, McGraw-Hill, New York, 1955.

5 R. Courant and D. Hilbert, Methods of Mathematical Physics, Vol.1, (first English edition), Interscience, New York, 1962.

6 R.W. Dickey, The Plane Circular Elastic Surface Under Normal Pressure. Arch for Ratl.Mechs. and Anal. 26 (1967), 219-236.

7 F.G. Tricomi, Integral Equations, Interscience, New York, 1957.

# 2 A static problem for the nonlinear string

The object of this lecture is to describe all possible equilibrium solutions of a string acted upon by a constant vertical force. If a string is in sufficiently high tension so that the linear theory applies, it is well known that the unique equilibrium position is a parabola. This linear theory is not valid when the tension in the string is small. For example, if the distance between the ends of the string is less than the actual unstressed length of the string, then a different theory is required. The most elementary such theory assumes the string is inextensible. In this case the equilibrium position of the string is given by those functions which make the integral

$$(2.1)\qquad \int_0^\ell y\sqrt{(1 + (y')^2)}\,dx$$

stationary subject to the condition that

$$(2.2)\qquad \int_0^\ell \sqrt{(1 + (y')^2)}\,dx = L$$

where it is assumed that the length of the string is $2L$ and the distance between the end points is $2\ell$. Assuming symmetry of deformation, the admissible functions must satisfy the boundary conditions $y'(0) = 0$ and $y(\ell) = a$. It follows that equilibrium solutions are given by those functions which satisfy

$$(2.3)\qquad \int_0^\ell y\sqrt{(1 + (y')^2)} + \lambda\sqrt{(1 + (y')^2)}\,dx = \text{STAT}.$$

The Euler equation for (2.3) is

$$(2.4)\qquad \frac{d}{dx}\frac{(y + \lambda)y'}{\sqrt{(1 + (y')^2)}} - \sqrt{(1 + (y')^2)} = 0,$$

or after simplification

$$(2.5)\qquad (y + \lambda)\,y'' - (1 + (y')^2) = 0$$

The solution of this equation satisfying the boundary condition $y' = 0$ at $x = 0$ is

(2.6) $$y = A \cosh \frac{x}{A} - \lambda.$$

It is also clear that for any choice of $A$, it is possible to choose $\lambda$ such that the boundary condition $x = 1$ is satisfied. The constant $A$ is determined by the requirement that the solution should satisfy (2.2). Thus there is a solution for every $A$ such that

(2.7) $$A \sinh \frac{\ell}{A} = L$$

The equation (2.7) can be rewritten

(2.8) $$\exp(\ell/A) = \frac{L}{A} + \sqrt{(\frac{L^2}{A^2} + 1)}.$$

It may be verified that if $L > \ell$, then the equ. (2.8) has not one, but three solutions: one with $A > 0$, and two with $A < 0$. These latter two cateneries correspond to equilibrium solutions in the form of arches (cf. fig. 2.2), and can be expected to be physically unstable. Nonetheless, they do exist as possible equilibrium solutions.

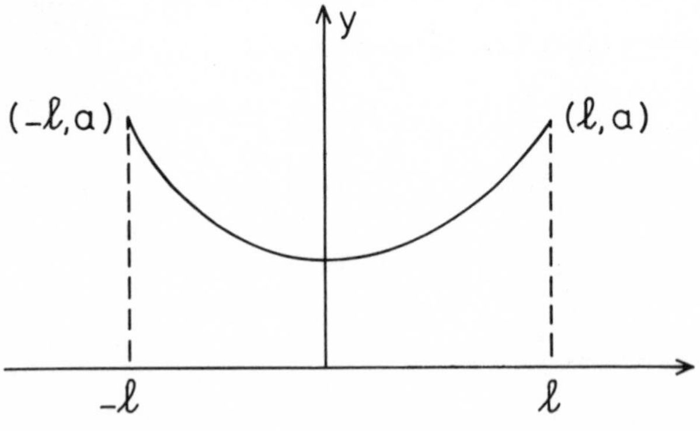


fig. 2.1

Although the preceding theory assumes the string is inelastic, similar results are obtained in the elastic case. An elementary elastic theory (due to Foppl) yields an equation of the form

$$(2.9) \qquad (c_0 + c_1 \int_{-\ell}^{\ell} (w')^2 dx) w'' = -p$$

where $w$ is the vertical displacement and should satisfy the boundary conditions $w'(0) = 0$ and $w(\ell) = 0$. The string is assumed to have length $2\ell$ and $c_0$ is positive or negative depending on whether the distance between the ends of the string is greater than or less than $2\ell$. The constant $c_1$ depends only on the parameters of the string and is always positive.

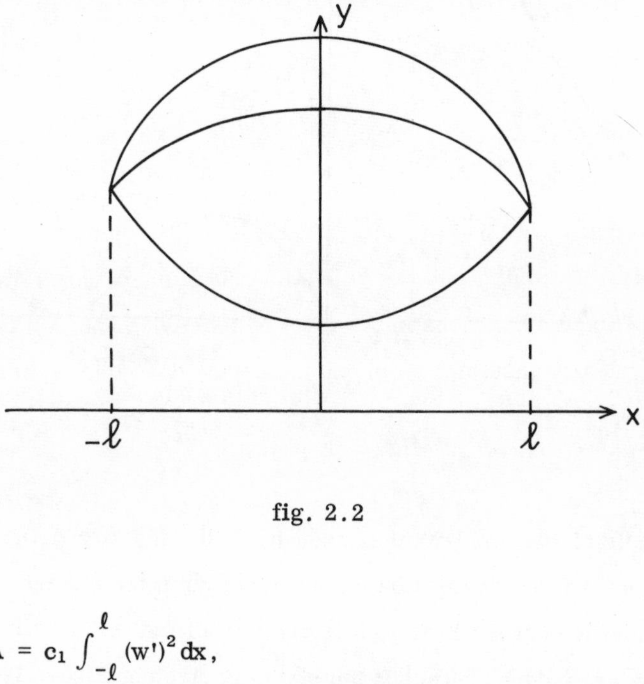


fig. 2.2

Let

$$(2.10) \qquad A = c_1 \int_{-\ell}^{\ell} (w')^2 dx,$$

and note that while $A$ depends on $w$, it is independent of $x$. Thus the solution of (2.9) satisfying the boundary conditions is given by

$$(2.11) \qquad w = \frac{p(\ell^2 - x^2)}{2(c_0 + A)},$$

where $A$ is to be determined from the equation (cf. (2.10))

$$(2.12) \qquad A = c_1 \int_{-\ell}^{\ell} \frac{p^2 x^2}{(c_0 + A)^2}\, dx$$

or equivalently

$$(2.13) \qquad F(A) = A(c_0 + A)^2 = \frac{2c_1 p^2 \ell^3}{3}$$

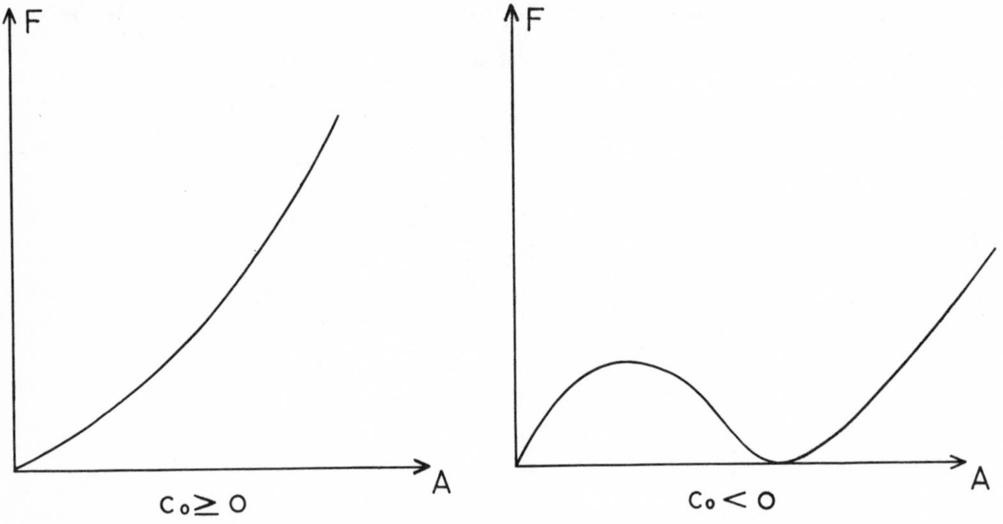


fig. 2.3

In fig. 2.3 $F(A)$ is plotted for both the case $c_0 \geq 0$ and $c_0 < 0$. The possible equilibrium solutions are determined by the intersection of the straight line $F = 2c_1 p^2 \ell^3/3$ with the curve $F(A)$. It is clear from fig. 2.3 that if $c_0 \geq 0$, these intersections are unique. Thus if the string is stretched there is only one solution. However, in the case in which the distance between the ends of the string is less than $2\ell$ and $p$ is sufficiently small, there are, just as in the theory described above, three possible solutions. The situation is described graphically in fig. 2.4.

The object now is to show that similar results can be expected even in the 'exact' theory of nonlinear strings. The equilibrium equations for a nonlinear string can be put in

$$(2.14) \qquad \frac{d}{d\xi}\, T\cos\theta = 0, \quad \frac{d}{d\xi}\, T\sin\theta = -p$$

where

$$(2.15) \qquad \cos\theta = \frac{1 + u_\xi}{1 + e}, \quad \sin\theta = \frac{1 + w_\xi}{1 + e}$$

and

$$(2.16) \qquad e = \sqrt{(1 + u_\xi)^2 + w_\xi^{\,2}} - 1.$$

The stress $T$ is related to the strain by a constitutive relation of the form

$$(2.17) \qquad T = f(e).$$

The quantities $u$ and $w$ are the dimensionless axial and vertical displacements, respectively. Assuming symmetric deformations, the boundary conditions are $u(0) = w_\xi(0) = w(1) = 0$ and $u(1) = \nu$, $-1 < \nu < \infty$. For simplicity it has been assumed that $p$ -- the dimensionless force -- is constant.

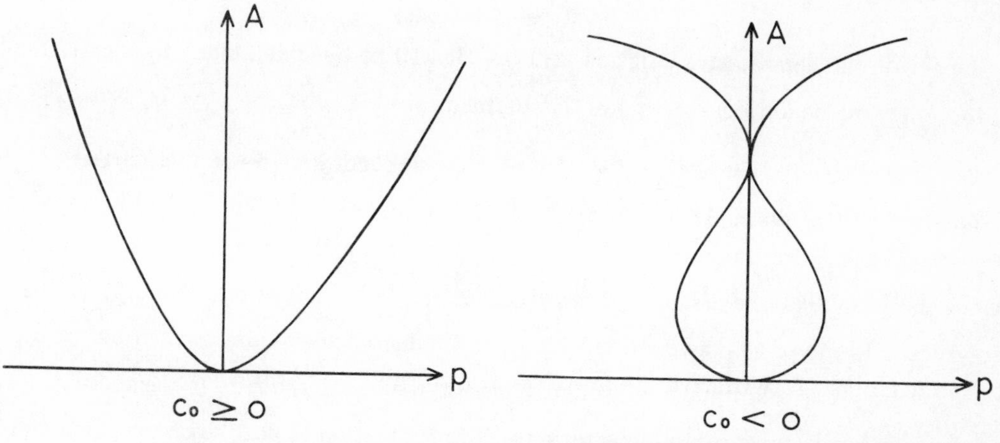


fig. 2.4

Equs. (2.14) can be integrated to show that $T$ has the form

(2.18) $T^2 = B^2 + p^2\xi^2$

where $B$ is a constant of integration. Thus the form of the stress is independent of the constitutive relation (2.17). In principle the problem is now solved since the other unknowns are given by quadratures of $T$. Thus

$$(2.19) \qquad w = \int_{\xi}^{1} \frac{(1+e)P}{T} \tau \, d\tau$$

and

$$(2.20) \qquad u + \xi = B\int_0^{\xi} \frac{1+e}{T} \, d\tau .$$

In addition, the boundary conditions are all satisfied with the possible exception of $u(1) = \nu$. This boundary condition will be satisfied if $B$ can be chosen so that

$$(2.21) \qquad \nu + 1 = B\int_0^{1} \frac{1+e}{T} \, d\tau .$$

The properties of

$$(2.22) \qquad I(B,P) = B\int_0^{1} \frac{1+e}{T} \, d\tau$$

depend on the constitutive relation (2.17). It will be assumed that $f(e)$ satisfies the following conditions: i) $f(e)$ is defined for $-1 \le e_- < e < e_+ \le \infty$; ii) $f(0) = 0$; iii) $0 < \frac{df}{de} < \infty$. Under these conditions there is a unique function $g(T)$ defined for $T_- < T < T_+$ where

$$(2.23) \qquad \lim_{e \to e_{+,-}} f(e) = T_{+,-} .$$

The function $g$ will have the properties: i) $g(0) = 0$; ii) $0 < \frac{dg}{dT} < \infty$; iii) $\operatorname{sgn} g(T) = \operatorname{sgn} T$; iv) $g(T) > -1$.

Under the above conditions $I(B,P)$ can be written

$$(2.24) \qquad I(B,P) = B\int_0^{1} \frac{1+g(T)}{T} \, d\tau .$$

For simplicity of analysis it will be assumed that $e_- = -1$, $e_+ = \infty$, $T_- = -\infty$, and $T_+ = +\infty$. Solutions of (2.21) for a fixed value of P are given by the intersections of the straight lines $I = \nu + 1$ with the curve $I = I(B,P)$. In order to describe these intersections, it is convenient to distinguish two cases.

Definition: T is tensile (compressive)if $T > 0$ ($T < 0$).

It is convenient to begin by discussing tensile solutions, i.e.,

$$T = +\sqrt{(B^2 + P^2\xi^2)}. \tag{2.25}$$

Theorem (2.1): The tensile solution (if it exists) is unique.

Proof: In order to prove this result, it suffices to show that I(B,P) is monotone increasing. (Note that since $\nu + 1 > 0$ and $T > 0$, there are no tensile solutions unless $B > 0$). After differentiation of I(B,P), it is found that

$$\frac{\partial I}{\partial B} = \int_0^1 \frac{B^2Tg' + P^2\xi^2(1+g)}{T^3} > 0. \tag{2.26}$$

Since the integrand of (2.26) is positive, I(B,P) is monotone increasing. Q.E.D.

Theorem (2.2): There exists a tensile solution for $-1 < \nu < \infty$.

Proof: Since it is already known that I(B,P) is a monotone increasing function of B, it suffices to show that

$$\lim_{B\to 0} I(B,P) = 0 \tag{2.27}$$

and

$$\lim_{B\to\infty} I(B,P) = \infty. \tag{2.28}$$

However,

$$(2.29)\quad I(B,P) \le B(1 + g(\sqrt{(B^2 + P^2)})) \int_0^1 \frac{dT}{\sqrt{(B^2 + P^2\tau^2)}}$$

$$= (1 + g(\sqrt{(B^2 + P^2)}) \frac{B}{P} \log \frac{P + \sqrt{(P^2 + B^2)}}{B}$$

and the right side of (2.29) goes to zero as $B \to 0$. Similarly,

$$(2.30)\quad I(B,P) \ge B(1 + g(B)) \int_0^1 \frac{dT}{\sqrt{(B^2 + P^2\tau^2)}}$$

$$= (1 + g(B)) \frac{B}{P} \log \frac{P + \sqrt{(P^2 + B^2)}}{B}$$

and the right side of (2.30) goes to infinity as $B \to \infty$. Q.E.D.

It is now of interest to consider compressive solutions. The first object is to show that there are no compressive solutions if $\nu \ge 0$.

Theorem (2.3): There are no compressive solutions if $\nu \ge 0$.

Proof: The proof is by contradiction. Assume there exists a compressive solution for $\nu \ge 0$. Then (2.21) has a solution for $\nu + 1 \ge 1$ and

$$(2.31)\quad T = -\sqrt{(B^2 + P^2\xi^2)}.$$

The mean value theorem implies the existence of a number $\xi^*$ such that $0 < \xi^* < 1$ and

$$(2.32)\quad \nu + 1 = B\frac{1 + g(T^*)}{T^*}$$

where

$$(2.33)\quad T^* = -\sqrt{(1 + P^2\xi^{*2})}.$$

It follows from (2.32) and (2.33) that

$$(2.34)\quad [\,(1 + g(T^*))^2 - (\nu + 1)^2]T^{*2} = (P\xi^*)^2(1 + g(T^*))^2.$$

The right side of (2.34) is positive. However, since $1 + g(T^*) < 1$ and

$\nu + 1 \geq 1$, the left side is negative, which is a contradiction. Q.E.D.
In view of theorem (2.3), it is only necessary to consider the case $-1 < \nu < 0$.

If $T < 0$, it cannot be expected that $I(B,P)$ is a monotone function of $B$. In fact, it is exactly this lack of monotonicity which introduces a multiplicity of compressive solutions when $P$ is sufficiently small. The object then is to determine the behaviour of $I(B,P)$ with $B$ and $T$ negative.

Lemma (2.1): $I(B,0) \to 1$ as $B \to 0$ and $I(B,0) \to 0$ as $B \to -\infty$.

Proof:

$$(2.35) \qquad I(B,0) = (1 + g(B)).$$

The quantity $(1 + g(B)) \to 1$ as $B \to 0$ and $1 + g(B) \to 0$ as $B \to -\infty$ since

$$(2.36) \qquad g(B) \to -1$$

as $B \to -\infty$. Q.E.D.
It may also be noted that

$$(2.37) \qquad \frac{\partial I(B,0)}{\partial B} = g'(B) > 0$$

so that $I(B,0)$ appears as in fig. 2.5

Lemma (2.2): If $B \neq 0$ then $\lim_{P \to 0} I(B,P) = I(B,0)$.

The proof follows on estimating the quantity $|I(B,P) - I(B,0)|$ when $B \neq 0$.

Lemma (2.3): If $P \neq 0$ then $\lim_{B \to 0} I(B,P) = 0$.

Proof:

$$(2.38) \qquad |I(B,P)| \leq |B|(1 + g(B)) \int_0^1 \frac{dT}{\sqrt{(B^2 + P^2\tau^2)}}$$

$$\leq (1 + g(B)) \frac{|B|}{P} \log \frac{P + \sqrt{(P^2 + B^2)}}{B}.$$

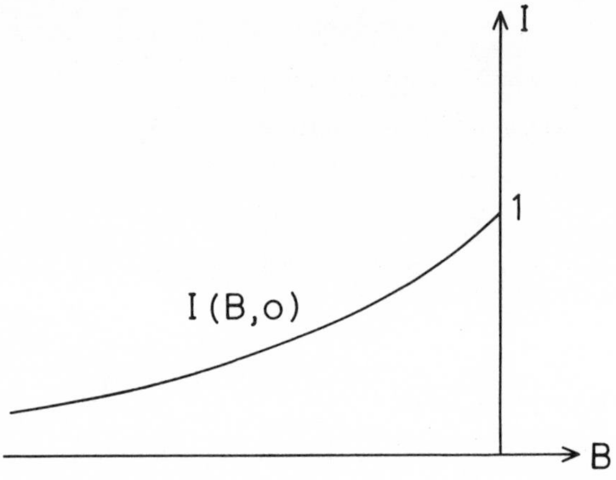


fig. 2.5

The right side of (2.38) approaches zero as $B \to 0$. Q.E.D.

It is a consequence of the preceding lemmas that $I(B,P)$ converges to $I(B,0)$ as long as $B \neq 0$. However, $I(B,P)$ always goes continuously to zero when $P \neq 0$. Thus the convergence of $I(B,P)$ to $I(B,0)$ cannot be uniform. The result must be as depicted in fig. 2.6.

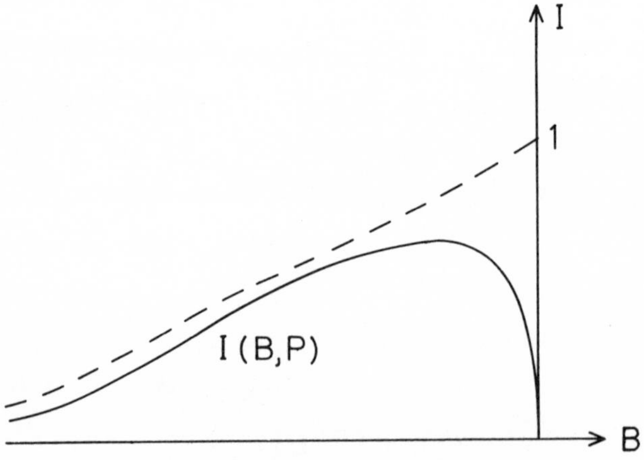


fig. 2.6

In view of fig. 2.6, if $P$ is sufficiently small the line $I = \nu + 1 \ (\nu < 0)$ must intersect the curve $I = I(B,P)$ at least twice.

Theorem (2.4): If $P$ is sufficiently small and $-1 < \nu < 0$, there are at least two compressive solutions.

The theorem states at least two compressive solutions since the situation depicted in fig. 2.7 has not been excluded.

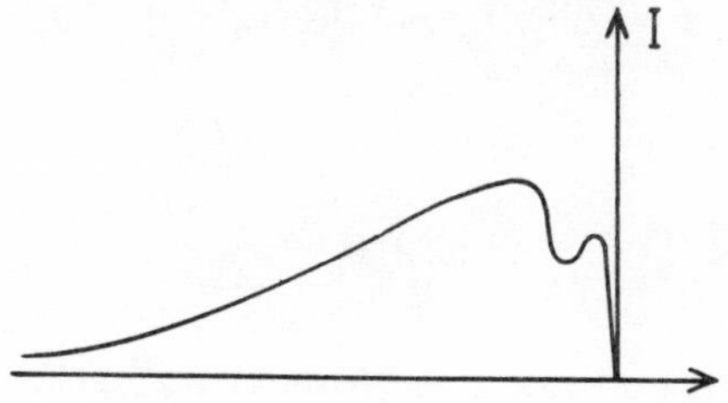


fig. 2.7

Theorem (2.5): Let $\nu$ be some fixed value such that $-1 < \nu < 0$. If $P$ is sufficiently large, there exist no compressive solutions.

In order to verify theorem (2.5), it suffices to show that $I(B,P) \to 0$ as $P \to \infty$.

The preceding results may be summarized as follows: i) For all $\nu$ such that $-1 < \nu < \infty$ and for all $P > 0$ there exists a unique tensile solution; ii) The tensile solution is the only solution if $\nu \geq 0$; iii) If $-1 < \nu < 0$ and $P$ is sufficiently small, there are at least two compressive solutions; and iv) If $-1 < \nu < 0$ and $P$ is sufficiently large, there are no compressive solutions.

The above results have been obtained under the assumption that $T_- = -\infty$ and $e_- = -1$. However, the other cases introduce little additional difficulty. The results are given in fig. 2.8. Thus the summary given above applies equally well to these cases.

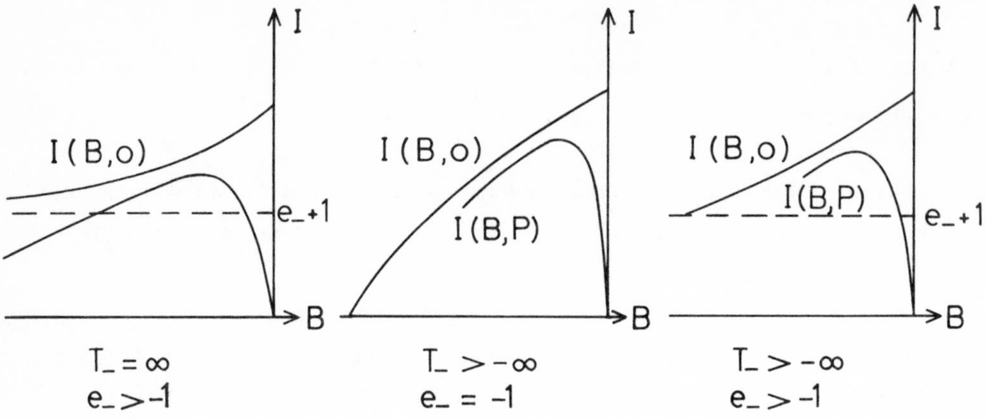


fig. 2.8

We conclude with the special example of linear materials, i.e.,

(2.39) $\quad T = Ee$

where the constant $E$ is Young's modulus. $g(T) = T/E$ is defined for $T > T_- = -E$. For these materials $I(B,P)$ can be explicitly evaluated as

(2.40) $$I(B,P) = \frac{B}{E} + \frac{|B|}{P} \log \frac{P + \sqrt{(P^2 + B^2)}}{B} .$$

The resulting curve is given in fig. 2.9.

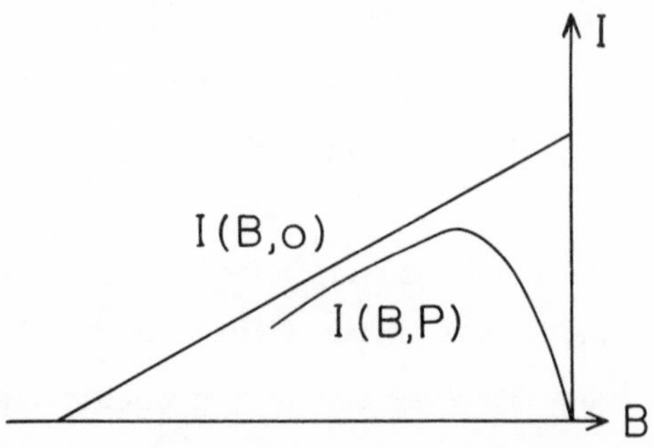


Fig. 2.9

The curve $I(B,P)$ terminates because the strain is defined only when $e > -1$. Thus when $P^2$ is greater than $E^2$, there are no compressive solutions.

References


1 R.W. Dickey, The Nonlinear String Under a Vertical Force, SIAM J. Appl. Math. 17 (1969), 172-178


Other References


2 G.F. Carrier, On The Nonlinear Vibration Problem of the Elastic String, Quart. Appl. Math. 3 (1945), 157-165.

# 3 A static problem for the nonlinear circular membrane

In the preceding chapter it was shown that under certain boundary conditions a nonlinear string has several equilibrium solutions. In this chapter it will be shown that a similar result holds for a circular membrane deformed by normal pressure. In fact, it will be shown that there exist situations in which infinitely many solutions are possible.

The equilibrium equations for a circular membrane acted on by a normal pressure $P(r)$ are given by

$$(3.1a)\qquad \frac{d}{dr}(r\sigma_r) = \sigma_\theta ,$$

$$(3.1b)\qquad \frac{d}{dr}\left(r\sigma \frac{dw}{rdr}\right) = -\frac{P(r)\,r}{h} ,$$

where $\sigma_r$ and $\sigma_\theta$ are the radial and circumferential stresses, $w$ is the normal displacement, and $h$ is the thickness of the surface. The strain-displacement equations are

$$(3.2a)\qquad \epsilon_r = \frac{du}{dr} + \frac{1}{2}\left(\frac{dw}{dr}\right)^2 ,$$

$$(3.2b)\qquad \epsilon_\theta = \frac{u}{r} ,$$

where $\epsilon_r$ and $\epsilon_\theta$ are the radial and circumferential strains, $u$ is the radial displacement. The stresses and strains are related to each other by Hooke's law, i.e.,

$$(3.3a)\qquad E\epsilon_r = \sigma_r - \nu\sigma_\theta ,$$

$$(3.3b)\qquad E\epsilon_r = \sigma_\theta - \nu\sigma_r ,$$

where $E$ is the Young's modulus and $\nu$ is the Poisson ratio. This set of six equations in six unknowns can be reduced to a single equation for the radial stress.

Thus if $x = r/R$ (R the radius of the unstrained surface) and

$$(3.4) \qquad y(x) = \frac{\sigma_r}{EP^{1/3}},$$

it can be shown that

$$(3.5) \qquad Ly = (x^3 y')' + G(x)/y^2 = 0, \qquad (' = d/dx)$$

where G depends on P and the other parameters. In the particular case where P is constant G is given by

$$(3.6) \qquad G = x^3.$$

If P does not vanish, then G satisfies the bounds

$$(3.7) \qquad kx^3 \leq G \leq Kx^3$$

where k and K are positive constants. The circumferential stress is related to y by

$$(3.8) \qquad \tau = (xy)'$$

where

$$(3.9) \qquad \tau = \sigma_\theta / EP^{1/3}.$$

The object is to discuss the existence of solutions to (3.5) under two distinct types of boundary conditions. For both types of boundary conditions symmetry will be assumed so that

$$(3.10) \qquad y'(0) = 0.$$

In addition, one other condition is needed. If stress is specified at the boundary (Problem T), the boundary condition at $x = 1$ is

$$(3.11) \qquad y(1) = T/EP^{1/3} = \lambda.$$

If radial displacement is specified at the boundary (Problem U), the boundary condition at $x = 1$ is

(3.12) $y'(1) + ay(1) = u(R)/RP^{1/3} = \mu$

where

(3.13) $1 - \nu = a > 0.$

In treating problems T and U, it is convenient to distinguish between different types of solutions.

Definition: $y$ is tensile (compressive) if
$y > 0$ $(y < 0)$ $0 \leq x \leq 1.$

Definition: $y$ is nodal if there exists $t$ such that
$0 < t < 1$ and $y(t) = 0.$

In the discussion below it will be shown that in certain circumstances problem T and problem U have multiple compressive solutions.

The method to be employed in solving the above-described boundary value problems is the so-called 'shooting' method, i.e., an initial value problem.

(3.14) $Lw = (x^3 w')' + G/w^2 = 0$

(3.15) $w(0,c) = c, w'(0,c) = 0$

is solved with the object of choosing C so as to solve the boundary value problem. It is not immediately obvious that this initial value problem has a solution since the equation is singular at $x = 0$, and hence the standard existence theorems do not apply. Nonetheless, it can be shown that:

Theorem (3.1): For every value of $c > 0$ and $c > s > 0$, there exists a unique $x_0(c,s)$ such that the initial value problem (3.14) and (3.15) has a unique solution in $0 \leq x \leq x_0$ with $w(x_0,c) = s$. Furthermore $w$ is monotone non-increasing for $0 < x < x_0$.

Theorem (3.2): For every value of $c < 0$, the initial value problem (3.14) and

(3.15) has a unique solution for all $x$ in the interval $0 \leq x < \infty$.

These two theorems are proven by using the method of successive approximation to show the existence of a solution close to $x = 0$, and then observing that the solution will continue to exist as long as $y$ does not vanish. However, since $y' < 0$ when $x > 0$ the result follows (cf. fig. 3.1).

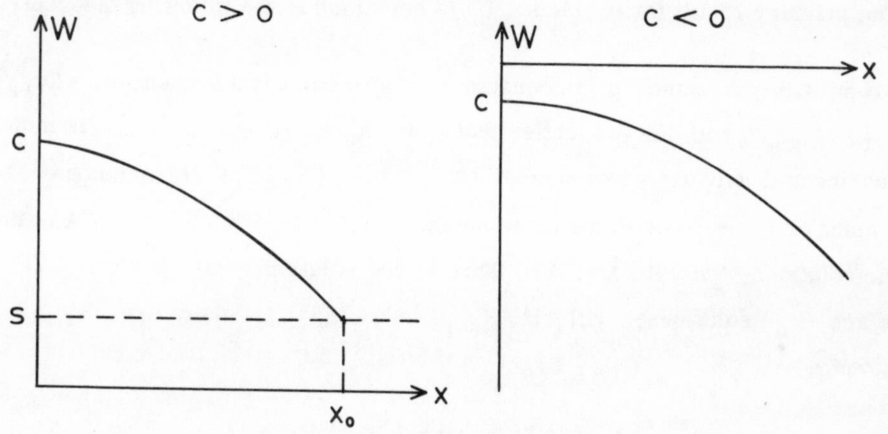


fig. 3.1

As has been observed, the solution of (3.14) and (3.15) is a solution of problem T if

(3.16) $\quad w(1,c) = \lambda .$

The solution of (3.14) and (3.15) is a solution of problem U if

(3.17) $\quad w'(1,c) + aw(1,c) = \mu .$

Theorem (3.3): If $p > 0$ and $\lambda < 0$, there are no tensile solutions.

Proof: Since $\lambda < 0$, the continuity of $y$ would imply $y < 0$ in a neighbourhood of $x = 1$. Q.E.D.

Theorem (3.4): If $p > 0$ there are no nodal solutions.

Proof: If $y(x)$ vanishes for some value of $x$ in the interval $0 < x < 1$, the differential equation (3.14) implies that either $y'$ or $y''$ is infinite there. However, by assumption $y'$ and $y''$ are continuous for $0 \leq x \leq 1$. Q.E.D.

The above two theorems exclude both tensile and nodal solutions for problem T when $p > 0$. However, problem U may have both compressive and tensile solutions when $\mu < 0$.

The primary result for problem T is contained in the following theorem:

Theorem (3.5): Assume $p$ is constant. There exist two sequences, $\{\lambda_{2n}\}$ and $\{\lambda_{2n+1}\}$, with the properties that: i) $\lambda_n < 0$; ii) $\{\lambda_{2n}\}$ is monotone decreasing and converges to a number $\lambda_\infty$; iii) $\{\lambda_{2n+1}\}$ is monotone increasing and converges to the same number $\lambda_\infty$; iv) if $0 > \lambda > \lambda_0$ there are no solutions; v) if $\lambda < \lambda_1$ there is one solution; vi) if $\lambda_{2n+2} < \lambda < \lambda_{2n}$ there are $2_n$ solutions; vii) if $\lambda_{2n+1} < \lambda < \lambda_{2n+3}$ there are $2_{n+3}$ solutions.

Proof: By defining a new dependent variable

$$(3.18) \qquad z = w/|c|$$

and a new independent variable

$$(3.19) \qquad t = \frac{x}{|c|^{3/2}}$$

the initial value problem (3.14) and (3.15) becomes (cf. (3.6))

$$(3.20) \qquad \frac{d}{dt}\left(t^3 \frac{dz}{dt}\right) + \frac{t^3}{z^2} = 0$$

$$(3.21) \qquad z(0) = \operatorname{sgn} c, \quad z'(0) = 0.$$

If $z(t)$ is a solution of (3.20) and (3.21) in the interval $0 \leq t \leq t_0$, then $w(x,c) = |c|\, z\left(\frac{x}{|c|^{3/2}}\right)$ is the solution of the initial value problem (3.14) and (3.15) in the interval $0 < x < |c|^{3/2} t_0$. The next step in the argument is to reduce (3.20) and (3.21) to a pair of first order ordinary differential equations. For this purpose define

$$(3.22) \qquad \xi = t^{-\frac{2}{3}} z$$

$$(3.23) \qquad \eta = t^{\frac{1}{3}} z.$$

The functions $\xi(t)$ and $\eta(t)$ satisfy the differential equations

$$(3.24) \qquad \frac{d\eta}{dt} = -(8\eta/3 + 1/\xi^2)/t$$

$$(3.25) \qquad \frac{d\xi}{dt} = (\eta - 2\xi/3)/t.$$

The initial conditions on $\xi$ and $\eta$ are determined from the fact that

$$(3.26) \qquad \lim_{t \to 0} t^{\frac{2}{3}} \xi(t) = \lim_{t \to 0} z(t) = \operatorname{sgn} c$$

$$(3.27) \qquad \lim_{t \to 0} t^{-\frac{1}{3}} \eta(t) = \lim_{t \to 0} z'(t) = 0$$

so that

$$(3.28) \qquad \lim_{t \to 0} \xi(t) = \infty \operatorname{sgn} c$$

$$(3.29) \qquad \lim_{t \to 0} \eta(t) = 0.$$

The equations (3.24) and (3.25) with initial conditions (3.28) and (3.29) are equivalent to the single first order equation

$$(3.30) \qquad \frac{d\eta}{d\xi} = \frac{-(8\eta/3 + 1/\xi^2)}{(\eta - 2\xi/3)}$$

with initial condition

$$(3.31) \qquad \eta(\infty \operatorname{sgn} c) = 0.$$

The equation (3.30) has one singular point given by the intersection of the curves $8\eta/3 + 1/\xi^2 = 0$ and $\eta - 2\xi/3 = 0$, i.e., at $\xi_0 = -(\frac{9}{16})^{\frac{1}{3}}$ and $\eta = -(\frac{1}{6})^{\frac{1}{3}}$. It may also be verified from the Bendixson theorem that equ. (3.30) has no limit cycles. Thus the orbit corresponding to (3.30) with initial data (3.31) is as depicted in fig. 3.2.

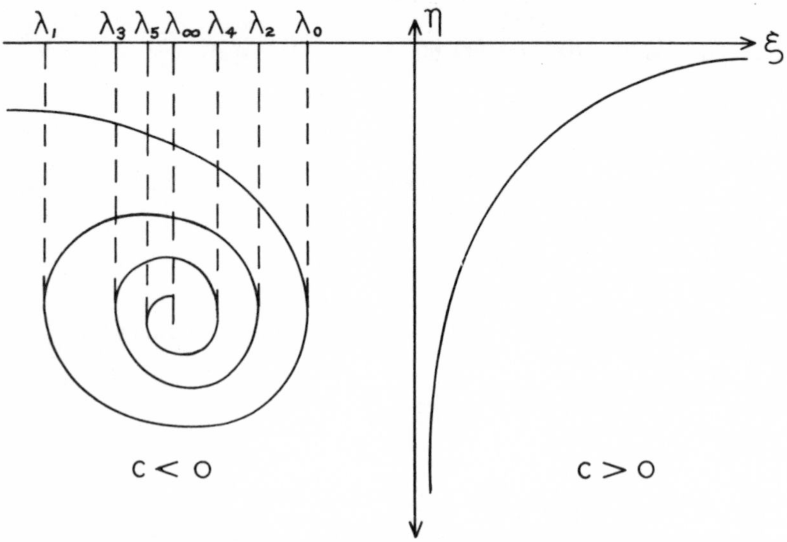


fig. 3.2

The points at which $|d\eta/d\xi| = \infty$ and $\xi > \xi_0$ are denoted by $\lambda_{2n}$, while the points for which $|d\eta/d\xi| = \infty$ and $\xi < \xi_0$ are denoted by $\lambda_{2n+1}$. It may be verified that if $\xi(\frac{1}{|c|^{3/2}}) = \lambda$ then $w(1,c) = \lambda$. It follows that every intersection of the line $\xi = \lambda$ with the spiral in fig. 3.2 corresponds to a compressive solution of problem T, since $|c|$ can be chosen arbitrarily. By simply counting the intersections of the line $\xi = \lambda$ with the spiral, the theorem follows. Note that if $\lambda = \lambda_\infty$ there are infinitely many intersections and hence infinitely many solutions. Q.E.D.

The preceding theorem gives a complete description of problem T with $\lambda < 0$. Problem U may be treated in a similar manner if P is constant. In fact, solutions of problem U are given by the intersections of the line $\eta + a\xi = \mu$

(cf. (3.17)) and the orbits in fig. 3.2 (cf fig. 3.3).

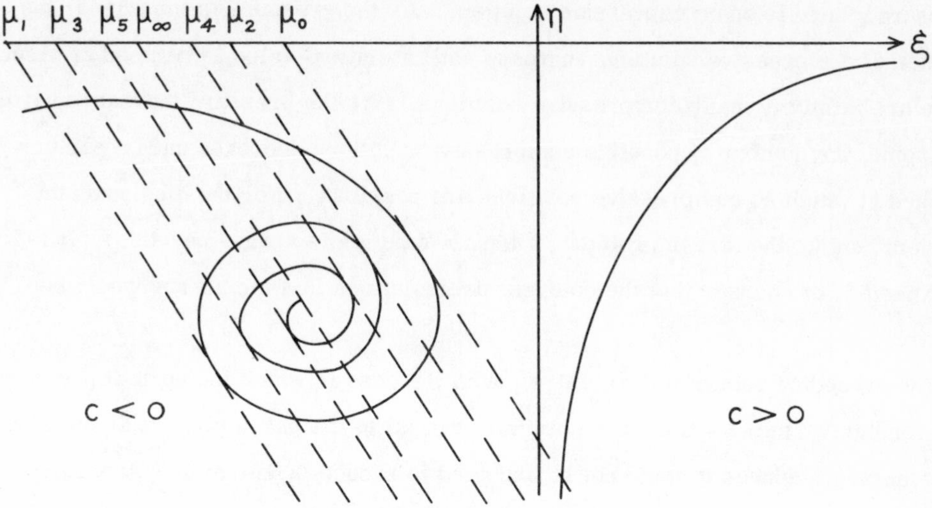


fig. 3.3

Theorem (3.6): Assume P is constant. There exist two sequences, $\{\mu_{2n}\}$ and $\{\mu_{2n+1}\}$, with the properties that: i) $\{\mu_{2n}\}$ is monotone decreasing and converges to a number $\mu_\infty$; ii) $\{\mu_{2n+1}\}$ is monotone increasing and converges to the same number $\mu_\infty$; iii) if $\mu > \mu_0$ there are no compressive solutions; iv) if $\mu < \mu_1$ there is one compressive solution; v) if $\mu_{2n+2} < \mu < \mu_{2n}$ there are $2_n$ compressive solutions; vi) if $\mu_{2n+1} < \mu < \mu_{2n+3}$ there are $2_{n+3}$ compressive solutions; vii) if $\mu = \mu_n$ there are $n + 1$ compressive solutions; and viii) in all cases there is a unique tensile solution.

The physical interpretation of these results is quite simple. If a compressive stress is prescribed on the boundary and the normal pressure is sufficiently small, there is exactly one compressive solution, and no other solutions are possible. As the pressure is increased, the number of possible compressive solutions also increase until $P = (T/E\lambda_\infty)^3$. From this point onward the number of possible solutions decrease until finally no solutions exist. Of course, if $\lambda > 0$ then the only possibility is a unique tensile solution. The situation is similar for the displacement problem. Thus for a fixed negative displacement and sufficiently small pressure, there is one compressive solution. As the pressure is increased, the number of compressive solutions increase until a critical value is reached at which there are infinitely many compressive solutions. As the pressure increases beyond this point, the number of possible compressive solutions decrease until a point is reached at which no compressive solutions are possible. For the displacement problem, unlike the stress problem, a tensile solution is always possible. It is to be expected, of course, that the compressive solutions in a membrane would be unstable.

The preceding remarks deal entirely with the case in which the normal pressure is a constant. In fact, this assumption is critical to the phase plane analysis since for general pressures it could not be expected to reduce (3.14) to an autonomous system of the form (3.20). Thus the results in this more general situation are much less precise. However, it is possible to show that there exists a value $\lambda = \lambda_0$ $(\mu = \mu_0)$ such that if $\lambda > \lambda_0$ $(\mu < \mu_0)$ there is at least one compressive solution.

Since, as has already been noted, $w(x,c)$ is a monotone decreasing function of $x$, it follows that $w(x,c) = \lambda$ $(\lambda, c < 0)$ has a unique solution if $0 > c > \lambda$. Denote this solution by $x_0(c,\lambda)$.

Lemma (3.1): $x_0(c,\lambda)$ is a continuous function of $c$.

Proof: The solution of the initial value problem (3.14) and (3.15) is a continuous function of $c$. In addition, $w'(x,c) < 0$. Thus the implicit function theorem implies the lemma. Q.E.D.

Lemma (3.2): $\lim_{c \to \lambda} x_0(c,\lambda) = 0$.

Proof: The differential equation (3.14) can be rewritten as a Volterra integral equation of the form

$$(3.32) \qquad w = c - \frac{1}{2} \int_0^x \left(\frac{1}{\eta^2} - \frac{1}{x^2}\right) \frac{G(\eta)}{w^2}\, d\eta .$$

Thus

$$(3.33)\qquad c - \lambda + \frac{1}{2}\int_0^{x_0} \left(\frac{1}{\eta^2} - \frac{1}{x_0^2}\right)\frac{G(\eta)}{w^2}\, d\eta .$$

The right side of (3.33) can be estimated by

$$(3.34)\qquad c - \lambda \geq \frac{k}{\lambda^2}\int_0^{x_0} \eta^3 \left(\frac{1}{\eta^2} - \frac{1}{x_0^2}\right) d\eta = \frac{k x_0^2}{2\lambda^2} .$$

It follows that $x_0 \to 0$ as $c \to \lambda$. Q.E.D.

Lemma (3.3): If problem T has a compressive solution for $\lambda = \lambda^*$, then it has a compressive solution for all $\lambda < \lambda^*$.

Proof: Let $c^*$ denote a solution of $w(1,c) = \lambda^*$. If $\lambda < \lambda^*$ the solution of $w(1,c) = \lambda^*$ has $x_0(c^*,\lambda) > 1$. (cf. fig. 3.4).

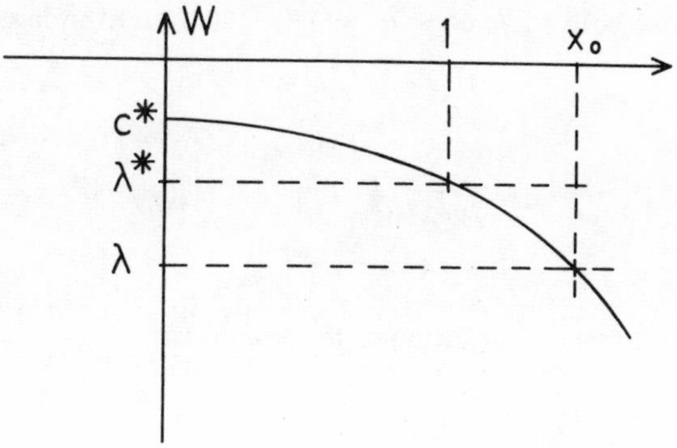


fig. 3.4

The lemma follows on application of lemma (3.2). Q.E.D.
In a similar manner it is shown that

Lemma (3.4): If problem T has no compressive solution for $\lambda = \lambda^*$, then there is no compressive solution for any $\lambda > \lambda^*$.

Theorem (3.7): There exists a critical value of $\lambda = \lambda_0$ such that there are no compressive solutions if $\lambda > \lambda_0$ and there are compressive solutions if $\lambda < \lambda_0$.

Proof: Assume there is a compressive solution to problem T when $\lambda = \lambda^*$. In this case equ. (3.32) implies that for some $c(0 > c > \lambda^*)$

$$(3.34) \quad c - \lambda^* \geq \frac{1}{2(\lambda^*)^2} \int_0^1 (\frac{1}{\eta^2} - 1)\, G(\eta)\, d\eta$$

so that

$$(3.35) \quad (\lambda^*)^3 \leq (\lambda^*)^3 - c(\lambda^*)^2 < \frac{1}{2} \int_0^1 (1 - \frac{1}{\eta^2})\, G(\eta)\, d\eta = \lambda_\mu^{\,3}.$$

Thus if $\lambda > \lambda_\mu$ there are no compressive solutions. On the other hand, if problem T has no compressive solution for $\lambda = \lambda^*$, then

$$(3.36) \quad c - \lambda^* = \frac{1}{2} \int_0^{x_0} (\frac{1}{\eta^2} - \frac{1}{x_0^2}) \frac{G(\eta)}{w^2}\, a\eta < \frac{1}{2} \int_0^1 (\frac{1}{\eta^2} - 1) \frac{G(\eta)}{w^2}\, d\eta$$

where the inequality must hold for $0 > c > \lambda^*$. The right side of (3.36) can be estimated by

$$(3.37) \quad \frac{1}{2} \int_0^1 (\frac{1}{\eta^2} - 1) \frac{G(\eta)}{w^2}\, d\eta \leq \frac{1}{2c^2} \int_0^1 (\frac{1}{\eta^2} - 1)\, G(\eta)\, d\eta = \frac{\lambda_\mu^{\,3}}{c^2} .$$

Thus if there are no compressive solutions, the inequality

$$(3.38) \quad c^2 (c - \lambda) < \lambda_\mu^{\,3}$$

must hold for $c = 2\lambda^*/3$ which is the value at which $c^2(c - \lambda)$ has its maximum. If this inequality is violated, there must be a compressive solution, i.e., if

$$(3.39) \quad \lambda^3 < (27/4)^{\frac{1}{3}} \lambda_\mu$$

there is a compressive solution. Q.E.D

It may be noted that the above theorem actually gives estimates on the critical value of $\lambda = \lambda_0$ for which compressive solutions exist. In fact, a better upper bound can be obtained by application of the Sturm theorem. It may be shown that $\lambda_0 \leq -3/(2j_{11})^{\frac{2}{3}}$ where $j_{11}$ is the smallest root of the Bessel function of first order.

The preceding remarks in this chapter have been devoted to a discussion of the existence of compressive solutions while saying little about their form.

Definition: $y_M(x)$ is a minimal solution of problem T if $y_M(x) \leq y(x)$ in the interval $0 < x < 1$ for all solutions $y(x)$ of problem T.

In a later chapter of these notes it will be shown that if problem T has a compressive solution, then it has a minimal compressive solution. At this juncture we will satisfy ourselves by simply assuming the existence of this solution and using it to derive other properties of compressive solutions.

Definition: Two functions, $y_1$ and $y_2$, are ordered on the interval $a \leq x \leq b$ if either $y_1 \geq y_2$ or $y_2 \geq y_1$ on that interval.

Theorem (3.8): Suppose for a fixed value of $\lambda$ there exist two distinct compressive solutions, $y_1$ and $y_2$, in addition to the minimal compressive solution $y_M$. Then $y_1$ and $y_2$ are not ordered.

Proof: Define

$$v(x) = y(x) - \lambda \tag{3.40}$$

where $y$ is a solution of problem T. Then $v$ will be a solution of

$$(x^3 v')' + \frac{G(x)}{(v + \lambda)^2} = 0 \tag{3.41}$$

and satisfy the boundary conditions

$$v(0) = v'(1) = 0. \tag{3.42}$$

The proof of the theorem is by contradiction. Assume $v_2 \geq v_1 \geq v_M$ where

$v_{1,2} = y_{1,2} - \lambda$. The functions

$$(3.43) \qquad f(x) = v_1 - v_M, \quad g(x) = v_2 - v_1$$

are positive in the interval $0 \leq x \leq 1$ and satisfy the equations

$$(3.44) \qquad (x^3 f')' + \psi(v_1, v_M) f = 0$$

$$(3.45) \qquad (x^3 g')' + \psi(v_1, v_2) g = 0$$

where

$$(3.46) \qquad \psi(\alpha, \beta) = -G(x) \left( \frac{\alpha^2 + \beta^2 + 2\lambda}{(\alpha+\lambda)^2 (\beta+\lambda)^2} \right).$$

In addition, the functions $f$ and $g$ satisfy the homogeneous boundary conditions (3.42). This fact can be rephrased as the equations:

$$(3.47) \qquad (x^3 f')' + \mu \psi(v_1, v_M) f = 0$$

$$(3.48) \qquad (x^3 g')' + \mu \psi(v_1, v_2) g = 0$$

satisfying the homogeneous boundary conditions have an eigenvalue $\mu = 1$. Moreover, in view of the fact that

$$(3.49) \qquad \psi(v_1, v_2) > \psi(v_1, v_M) > 0$$

and that both $f$ and $g$ are positive, $\mu = 1$ is the minimum eigenvalue (consequence of the Sturm theorem). We now show that this is not possible. It is well known that

$$(3.50) \qquad \min_{f \neq 0} \int_0^1 x^3 (f')^2 dx \Big/ \int_0^1 \psi(v_1, v_M) f^2 dx$$

yields the minimum eigenvalue. Thus if $f_0$ is the eigenfunction corresponding to $\mu = 1$

$$(3.51)\quad 1 = \int_0^1 x^3 (f_0')^2 dx / \int_0^1 \psi(v_{(.)} v_M) f_0 dx$$

$$> \int_0^1 x^3 (f_0')^2 dx / \int_0^1 \psi(v_1, v_2) f_0^2 dx$$

$$> \min_{g \not\equiv 0} \int_0^1 x^3 (g')^2 dx / \int_0^1 \psi(v_1, v_2) g^2 dx = 1.$$

This contradiction proves the theorem. Q.E.D.

It is the consequence of the above theorem that all compressive solutions, with the exception of the minimal compressive solution, must intersect each other.

References

1 A.J. Callegari, H.B. Keller, and E.L.Reiss, Membrane Buckling : A Study of Solution Multiplicity. Comm. on Pure and Appl.Math. 24 (1971), 499-521.

Other References

2 A.J. Callegari and E.L. Reiss, Nonlinear Boundary Value Problems for the Circular Membrane. Arch. for Ratl.Mech. and Anal. 31 (1960), 390-400.

3 R.W. Dickey, The Plane Circular Elastic Surface Under Normal Pressure. Arch. for Ratl.Mech. and Anal. 26 (1967) 219-236.

4 H B. Keller and E.L. Reiss, Iterative Solutions for the Nonlinear Bending of Circular Plates. Comm. on Pure and Appl.Math. 11 (1958), 273-292.

# 4 The rotating string

In the preceding two chapters static problems for nonlinear strings and membranes have been studied. In this section we treat a dynamic problem for the nonlinear string. In particular, we study the motion of a string of length $\ell$, fixed at one end and free to rotate about the $z$ axis (cf. fig. 4.1).

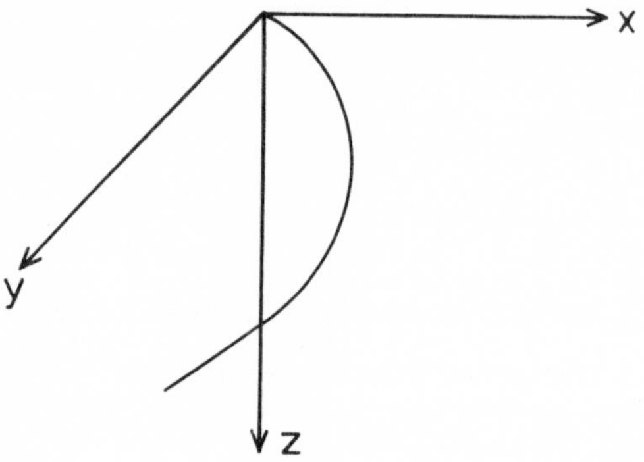


fig. 4.1

It is assumed that the only force acting on the string is the force of gravity $\bar{g} = (0,0,g)$. If the string is inextensible, the equations of motion can be written

$$(4.1) \qquad (\rho \bar{x}_t)_t = \rho \bar{g} + \frac{\partial}{\partial s}\left(T \frac{\partial \bar{x}}{\partial s}\right)$$

where $\bar{x} = (x,y,z)$ is the position vector, $\rho$ is the mass per unit length, and $T$ is the tension in the string. The independent variable $s$ is the arc-length measured from the free end point. Since the string is assumed inextensible, we have in addition to equ. (4.1) the additional constraint

$$(4.2) \qquad x_s^{\,2} + y_s^{\,2} + z_s^{\,2} = 1.$$

The boundary conditions to be imposed on equs. (4.1) and (4.2) are

(4.3a) $x(\ell,t) = y(\ell,t) = z(\ell,t) = 0,$

(4.3b) $T(0,t) = 0.$

It is not our purpose to study the solutions of (4.1) and (4.2) under arbitrary initial data, but rather to search for solutions of a very particular type. Thus we consider the question of whether equs. (4.1) and (4.2) have solutions of the form

(4.4) $x(s,t) + iy(s,t) = x(s)e^{iwt}$

(4.5) $z(s,t) = z(s)$

(4.6) $T(s,t) = T(s).$

If (4.4), (4.5) and (4.6) are solutions of equs. (4.1) and (4.2), then $x(s), z(s)$, and $T(s)$ must satisfy the ordinary differential equations

(4.7) $(Tx')' + \rho\omega^2 x = 0 \quad (' = d/ds)$

(4.8) $(Tz')' + \rho g = 0$

(4.9) $(x')^2 + (y')^2 = 1$

and boundary conditions

(4.10) $T(0) = x(\ell) = z(\ell) = 0.$

These three ordinary differential equations can be reduced to a single second order differential equation for the variable

(4.11) $u = Tx'/\rho g.$

In order to see this, simply note that equ. (4.8) and the boundary conditions (4.10) imply that

$$(4.12)\qquad Tz' = \rho gs.$$

Combining this result with equ. (4.9) and (4.11) yields

$$(4.13)\qquad T^2((x')^2 + (y')^2) = T^2 = \rho^2 g^2 (u^2 + s^2)$$

or equivalently

$$(4.14)\qquad T = +\rho g\sqrt{(u^2 + s^2)}$$

where the plus sign is chosen, since it is to be expected that the string will be in tension. Once $u$ is determined, $T$ is given explicitly by equ. (4.14). In addition, $x$ and $z$ are given by

$$(4.15)\qquad x = -\int_s^{\ell} u(u^2 + \xi^2)^{-\frac{1}{2}} d\xi,$$

$$(4.16)\qquad z = \int_s^{\ell} \xi(u^2 + \xi^2)^{-\frac{1}{2}} d\xi.$$

Thus it only remains to find $u$. After differentiating equ. (4.7), it follows easily from equ. (4.15) that $u$ must be a solution of

$$(4.17)\qquad u'' + \frac{\omega^2}{g}\frac{u}{\sqrt{(u^2 + s^2)}} = 0,$$

and satisfy the boundary conditions

$$(4.18)\qquad u(0) = u'(\ell) = 0.$$

By defining new variables

$$(4.19)\qquad \tilde{s} = s/\ell,\quad \tilde{u} = u/\ell,\quad \lambda = \omega^2 \ell/g$$

the problem is reduced to solving

$$(4.20)\qquad \tilde{u}'' + \lambda\frac{\tilde{u}}{\sqrt{(\tilde{u}^2 + \tilde{s}^2)}} = 0$$

with boundary conditions

(4.21) $\quad \tilde{u}(0) = \tilde{u}(1) = 0.$

If it is assumed that $\tilde{u}$ is small, then equ. (4.20) could be reasonably replaced by the equation

(4.22) $\quad \tilde{u}'' + \lambda \frac{\tilde{u}}{\tilde{s}} = 0.$

The solution of this equation satisfying the boundary conditions (4.21) is $\tilde{u} \equiv 0$ unless $\lambda = (\sigma_n/2)^2$ where $\sigma_n$ is a root of the Bessel function of zeroth order. In this case equ. (4.20) has a nontrivial solution

(4.23) $\quad \tilde{u} = \sqrt{(\tilde{s})} J_1(2\sqrt{(\lambda)}\tilde{s})$

where $J_1$ is the Bessel function of first order. Thus the linear theory would indicate that the string has nontrivial solutions of the desired form only in the case that the angular velocity has certain critical values.

It is, unfortunately, not clear that this linear theory would have any relation to the nonlinear theory, since it is not to be expected that $\tilde{u}$ would be small. Thus it is conceivable that the nonlinear problem might have nontrivial solutions for all $\lambda > 0$. However, we shall show that this is not possible. For this purpose simply note that equ. (4.20) with boundary conditions (4.21) can be rewritten in the form of a nonlinear integral equation

(4.24) $$\tilde{u}(\tilde{s}) = \int_0^1 G(\tilde{s},\xi) \frac{\tilde{u}}{\sqrt{(\tilde{u}^2 + \xi^2)}} d\xi, \quad G(\tilde{s},\xi) = \begin{cases} \xi & 0 \le \xi \le s \\ s & s \le \xi \le 1 \end{cases}.$$

Equation (4.24) implies that

(4.25) $$|\tilde{u}(\tilde{s})| \le \lambda \int_0^1 G(\tilde{s},\xi) \frac{|\tilde{u}|}{\xi} d\xi \le \lambda \left( \int_0^1 \frac{G(\tilde{s},\xi)^2}{\xi^2} d\xi \right)^{\frac{1}{2}} \|u\|$$

where

(4.26) $$\|u\| = \left(\int_0^1 u(\xi)^2 d\xi\right)^{\frac{1}{2}}$$

(4.27) $$\int_0^1 \frac{G(\tilde{s},\xi)^2}{\xi} d\xi = 2\tilde{s} - \tilde{s}^2.$$

Thus squaring both sides of equ. (4.25) and integrating with respect to $\tilde{s}$, we find

(4.28) $$\|u\|^2 \leq \tfrac{2}{3}\lambda^2 \|u\|^2.$$

If $u \neq 0$, then the inequality (4.28) implies that

(4.29) $$\lambda^2 \geq 3/2.$$

Thus equ. (4.20) has no nontrivial solution unless $\lambda$ satisfies the inequality (4.29). Actually a better bound on $\lambda$ can be attained. It can be shown that there are no nontrivial solutions unless $\lambda > (\sigma_1/2)^2$ where $\sigma_1$ is the first zero of the Bessel function of zeroth order. This latter result is the best possible, since the following theorem will be proven.

Theorem (4.1): Equ. (4.20) with boundary conditions (4.21) has for any $\lambda_n < \lambda < \lambda_{n+1}$ ($\lambda_n = (\sigma_n/2)^2$) exactly $n$ nontrivial solutions $\tilde{u}_1, \ldots, \tilde{u}_n$.

The function $\tilde{u}_\nu$ has exactly $\nu$ isolated zeros (including $\tilde{s} = 0$). It may be noted that if $\tilde{u}$ is a solution of equ. (4.20), then $-\tilde{u}$ is also a solution. In the above theorem the pair $(\tilde{u}, -\tilde{u})$ is counted as one solution.

Theorem (4.1) will be proved by obtaining certain results on the initial value problem

(4.30) $$v'' + \frac{v}{\sqrt{(v^2 + x^2)}} = 0$$

(4.31) $$v(0,a) = 0, \quad v'(0,a) = a > 0.$$

Theorem (4.2): (1) $v(x)$ exists, is unique, and depends continuously on $a$.

(2) $v(x)$ has an infinite number of isolated zeros $0 = y_0 < y_1 < \ldots < y_n < \ldots$ such that $y_n(a) \to \infty$ as $n \to \infty$. $v'(x)$ has

an infinite number of isolated zeros $z_1 < z_2 < \ldots < z_n < \ldots$ interlacing the zeros of $v(x)$, i.e., $y_0 < z_1 < y_1 < z_2 < y_2 < \ldots$. Furthermore $\lim_{a\to\infty} z_n(a) = \infty$ and $\lim_{a\to 0} z_n(a) = \lambda_n$ where $\lambda_n = (\sigma_n/2)^2$.

(3) $z_n(a)$ are differentiable functions of $a$ and $dz_n/da \geq 0$.

Theorem (4.2) implies that even though $v$ goes to zero as $a \to 0$, the zeros of $v$ go monotonically to the points $\lambda_n$ (cf. fig. 4.2).

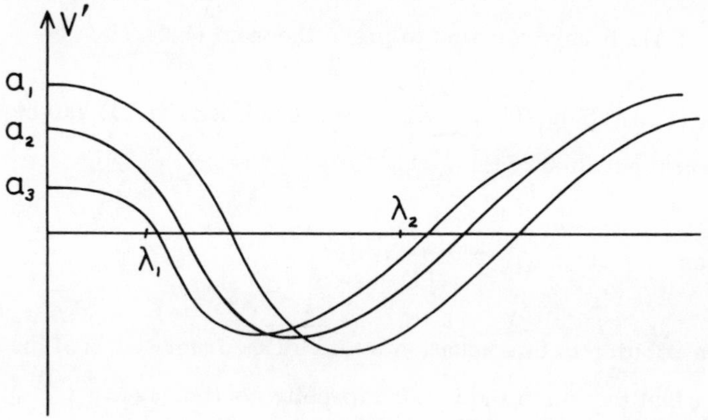


fig. 4.2

Lemma (4.1): Theorem (4.2) implies theorem (4.1).

Proof: We assume theorem (4.2) has been proved. Let $\lambda$ be any value such that $\lambda_n < \lambda < \lambda_{n+1}$. Since $z_n(a)$ is a monotone function and $z_n(a) \to \lambda_n$ as $a \to 0$, it follows that $z_n(a) < \lambda$ when $a$ is sufficiently small. Moreover, $z_\nu > \lambda_n$ for all $\nu > n$ and $a > 0$. Since $\lim_{a\to\infty} z_n(a) = \infty$ there exists a value of $a$, say $a_n(\lambda)$, such that $z_n(a_n(\lambda)) = \lambda$. The function $v(x, a_n)$ has the property that $v(0, a_n) = 0$ and $v'(\lambda, a_n) = 0$. Furthermore, $v(x, a_n)$ has zeros at $0 = y_0 < y_1 < \ldots < y_{n-1}$, i.e., $v(x, a_n)$ has $n$ zeros. Denote this function by $v_n$. If $a$ continues to increase, the fact that $z_{n-1}(a) \to \infty$ implies that there exists a value of $a$, say $a_{n-1}(\lambda)$, such that $z_{n-1}(a_{n-1}(\lambda)) = \lambda$. The function $v(x, a_{n-1})$ has the property that $v(0, a_{n-1}) = 0$ and $v(\lambda, a_{n-1}) = 0$.

Furthermore, $v(x, a_{n-1})$ has zeros at $0 = y_0 < \ldots < y_{n-2}$, i.e., $v(x, a_{n-1})$ has $n-1$ zeros. Denote this function by $v_{n-1}$. In this manner we obtain $n$ functions $v_1, \ldots, v_n$ where $v_\nu$ has $\nu$ isolated zeros. No other solutions have the property that $a > 0$ and $v'(\lambda) = 0$. It is easily verified that the functions

$$u_\nu(\tilde{s}) = \frac{1}{\lambda} v_\nu(\lambda \tilde{s})$$

are solutions of the boundary value problem (4.20) and (4.21). Q.E.D.

In view of lemma (4.1), it only remains to prove theorem (4.2).

Proof of Theorem (4.2): Equ. (4.30) with initial conditions (4.31) can be rewritten as a Volterra integral equation of the form

$$(4.32) \qquad v = ax - \int_0^x (x - \xi) \frac{v}{(v^2 + \xi^2)^{\frac{1}{2}}} \, d\xi .$$

The existence of a solution to this equation would follow from standard theorems except for the fact that the integrand is not Lipschitz continuous at $\xi = 0$. However, this difficulty is removed by defining a new variable

$$(4.33) \qquad v = xw$$

so that the equation for $w$ is given by

$$(4.34) \qquad w = a - \int_0^x (1 - \xi/x) \frac{w}{(w^2 + 1)^{\frac{1}{2}}} \, d\xi .$$

The integrand in equ. (4.34) is a Lipschitz continuous. In fact, defining

$$(4.35) \qquad D(y) = \frac{1}{(y^2 + 1)^{\frac{1}{2}}}$$

it follows that

$$(4.36) \qquad \frac{d}{dy} y D(y) = D(y)^3$$

and thus

(4.37) $$0 \leq \frac{d}{dy}\, yD(y) \leq 1.$$

The mean value theorem in conjunction with (4.37) implies that

(4.38) $$|\alpha D(\alpha) - \beta D(\beta)| \leq |\alpha - \beta|.$$

This result is sufficient to guarantee the global existence of a solution to (4.34) for $x \geq 0$.

Similarly, uniqueness is easily shown. Assume $w$ and $\tilde{w}$ are solutions of equ. (4.34). Then

(4.39) $$|w - \tilde{w}| \leq \int_0^x (1 - \xi/x)\, |wD(w) - \tilde{w}D(\tilde{w})|\, d\xi$$

$$\leq \int_0^x (1 - \xi/x)\, |w - \tilde{w}|\, d\xi.$$

Define

(4.40) $$\|w - w\|_\mu = \max_{0 \leq x \leq 1} |e^{-\mu x}(w - \tilde{w})|$$

so that

(4.41) $$|e^{-2x}(w - \tilde{w})| \leq e^{-2x} \int_0^x (1 - \xi/x)\, e^{2\xi}\, |e^{-2\xi}(w - \tilde{w})|\, d\xi$$

$$\leq \|w - \tilde{w}\|_2\, e^{-2x} \int_0^x e^{2\xi}\, d\xi.$$

The inequality (4.41) yields

(4.42) $$\|w - \tilde{w}\|_2 \leq \tfrac{1}{2} \|w - \tilde{w}\|_2$$

or equivalently

(4.43) $$w = \tilde{w}$$

which completes the proof of uniqueness.

In order to prove the continuous dependence of the solution on $a$, define

$$(4.44) \qquad \Delta w = w(x, a + \Delta a) - w(x, a)$$

so that $\Delta w$ satisfies the equation (cf. (4.38))

$$(4.45) \qquad \Delta w = \Delta a - \int_0^x (1 - \xi/x)\,[(w + \Delta w)D(w + \Delta w) - wD(w)]\,d\xi$$

$$\leq \Delta a - \int_0^x (1 - \xi/x)\,|\Delta w|\,d\xi .$$

Proceeding as above, we obtain the inequality

$$(4.46) \qquad |e^{-2x}\,\Delta w| \leq \Delta a e^{-2x} + \frac{1}{2}\,\|\Delta w\|_2 \leq \Delta a + \frac{1}{2}\,\|\Delta w\|_2$$

from which it follows that

$$(4.47) \qquad \|\Delta w\|_2 \leq 2\Delta a .$$

This completes the proof of part 1 of theorem (4.2).

It should be noted that it is a consequence of part 1 of theorem (4.2) that the zeros of $v$ and $v'$ (if they exist) depend continuously on $a$. The continuous dependence of the zeros of $v$ on $a$ is obvious. The continuous dependence of the zeros of $v'$ on $a$ follows from the fact that

$$(4.48) \qquad v'(x) = a - \int_0^x \frac{v}{(v^2 + \xi^2)^{\frac{1}{2}}}\,d\xi$$

and hence $v'$ depends continuously on $a$.

Although we have shown that $v$ is a continuous function of $a$, more will be required. In fact, we wish to show that $v$ is a differentiable function of $a$. For this purpose consider the equation

$$(4.49) \qquad (\delta v)'' + \frac{\delta v}{(v^2 + x^2)^{3/2}} = 0$$

with initial conditions

(4.50) $\delta v(0,a) = 0, \quad (\delta v(0,a))' = 1$

which is obtained by formally differentiating equ. (4.30) and the initial conditions (4.31) with respect to a. Let

(4.51) $\delta v = \Delta(x,a)$

be a solution of this equation satisfying the initial conditions (4.50). The existence of this solution follows from arguments essentially identical to those given above for equ. (4.30). Rewriting equ. (4.49) it follows that $\Delta$ is a solution of the equation

$$(4.52) \quad \Delta(x,a) = x - \int_0^x (x - \xi) \frac{\xi^2 \Delta}{(v^2 + \xi^2)^{3/2}} d\xi .$$

In order to remove the difficulty at $\xi = 0$, define

(4.53) $\Delta = x\,\delta w$

so that $\delta w$ satisfies the equation

$$(4.54 \quad \delta w = 1 - \int_0^x (1 - \xi/x) \frac{\delta w}{(w^2 + 1)^{3/2}} d\xi .$$

We now show that

$$(4.55) \quad \lim_{\Delta a \to 0} \frac{\Delta w}{\Delta a} = \frac{dw}{da} = \delta w$$

which is equivalent to showing that (cf. (4.33), (4.44))

$$(4.56) \quad \lim_{\Delta a \to 0} \frac{\Delta v}{\Delta a} = \frac{dv}{da} = \Delta .$$

Defining

$$(4.57) \quad d = \frac{\Delta w}{\Delta a} - \delta w$$

it follows from equ. (4.45) and equ. (4.54) that

$$(4.58) \quad d = -\int_0^x (1 - \xi/x)\left\{\frac{(w+\Delta w)\,D(w+\Delta w) - w\,D(w)}{\Delta a} - D^3(w)\,\delta w\right\} d\xi .$$

For the present purposes a better estimate on $(w+\Delta w)\,D(w+\Delta w) - w\,D(w)$ is needed than that furnished by (4.38). Noting that

$$(4.59) \quad \frac{d^2}{dy^2}\, y\,D(y) = -3\,y\,D^5(y)$$

so that

$$(4.60) \quad \left|\frac{d^2}{dy^2}\, y\,D(y)\right| \le 3.$$

Taylor's theorem with remainder implies that

$$(4.61) \quad |\alpha D(\alpha) - \beta D(\beta)| \le (\alpha - \beta)\,D^3(\beta) + |A|\,(\alpha - \beta)^2$$

where

$$(4.62) \quad |A| \le 3/2.$$

Placing this estimate in (4.58) we find that

$$(4.63) \quad |d| \le \int_0^x (1 - \xi/x)\left(|d|\,D^3(w) + \frac{3\,|\Delta w|^2}{2\Delta a}\right) d\xi .$$

Since $D^3(w) \le 1$, the above inequality can be replaced by

$$(4.64) \quad |d| \le \int_0^x (1 - \xi/x)\left(|d| + \frac{3\,|\Delta w|^2}{2\Delta a}\right) d\xi .$$

Multiplying (4.67) by exp(-4x), we find that

$$(4.65) \quad |e^{-4x}\,d| \le e^{-4x}\int_0^x (1 - \xi/x)\,e^{-4\xi}\left(|e^{-4\xi}\,d| + \frac{3\,|e^{-2\xi}\,\Delta w|^2}{2\Delta a}\right) d\xi .$$

Combining this result with (4.47) yields

$$(4.66) \quad \|d\|_4 \le 2\Delta a$$

which proves (4.56).

We now turn to the question of the existence of zeros of $v$ and $v'$. We first note that in view of the fact that

$$(4.67) \qquad 0 \leq \frac{d}{dx}\left(\frac{1}{2}(v')^2 + (v^2 + x^2)^{\frac{1}{2}}\right) \leq 1$$

we obtain the bounds

$$(4.68) \qquad 0 \leq \frac{1}{2}((v')^2 - a^2) + (v^2 + x^2)^{\frac{1}{2}} \leq x.$$

Thus

$$(4.69) \qquad |v'| \leq a,$$

$$(4.70) \qquad |v| \leq ax,$$

and

$$(4.71) \qquad |w| \leq a.$$

These bounds will allow us to use the Sturm comparison theorem to prove that $v$ has infinitely many isolated zeros. In fact, consider the two equations

$$(4.72) \qquad v'' + q(x)v = 0,$$

$$(4.73) \qquad y'' + \bar{q}\, y = 0$$

where (cf. (4.70))

$$(4.74) \qquad q(x) = \frac{1}{(v^2 + x^2)^{\frac{1}{2}}} \geq \frac{1}{(a^2x^2 + x^2)^{\frac{1}{2}}} = \frac{1}{x(a^2 + 1)^{\frac{1}{2}}} = \bar{q}\,.$$

The Sturm theorem implies that $v$ has at least one zero between consecutive zeros of $y$. However, a solution of (4.73) is

$$(4.75) \qquad y = (x)^{\frac{1}{2}} J_1 ((a^2 + 1)^{\frac{1}{4}} x^{\frac{1}{2}})$$

which has infinitely many isolated zeros. Thus $v$ has infinitely many isolated zeros $0 = y_0 < y_1 < \ldots < y_n < \ldots$. (Note that the zeros are isolated since if there were a limit point of zeros it would imply that the solution is identically zero.) The fact that $v'$ has infinitely many zeros follows from Rolle's theorem which implies that between each of the points $y_n$ and $y_{n+1}$ there exists at least one point, say $z_{n+1}$, such that $v' = 0$. The fact that there is only one zero of $v'$ between each of the points $y_n$ and $y_{n+1}$ follows from the fact that $v''$ does not change sign in this interval (cf. (4.30)). Note that in a similar way it can be shown that $\Delta$ and $\Delta'$ have infinitely many zeros interlacing each other. Thus if $\alpha_n$ are the zeros of $\Delta$ and $\beta_n$ are the zeros of $\Delta'$, then $0 = \alpha_0 < \beta_1 < \alpha_1 < \ldots.$

In order to complete the proof of part 2 of theorem (4.2), it remains to show that $z_n(a) \to \infty$ as $a \to \infty$ and $z_n(a) \to \lambda_n$ as $a \to 0$. In order to show that $z_n(a) \to \infty$ as $a \to \infty$, it suffices to show that $z_1(a) \to \infty$ as $a \to \infty$. However, since

$$\text{(4.76)} \qquad \frac{v}{(v^2 + x^2)^{\frac{1}{2}}} \leq 1$$

equ. (4.48) implies

$$\text{(4.77)} \qquad v'(x) \geq a - \int_0^x d\xi = a - x.$$

Thus

$$\text{(4.78)} \qquad z_1(a) \geq a$$

from which it follows that $z_1(a) \to \infty$ as $a \to \infty$. The major difficulty showing that $z_n(a) \to \lambda_n$ as $a \to 0$ is that $v \to 0$ as $a \to 0$. In order to remove this difficulty we consider the function $v/a$ rather than $v$ where we note that both $v/a$ and $(v/a)'$ have the same zeros as $v$ and $v'$. The object is to show that

$$\text{(4.79)} \qquad \lim_{a \to 0} \frac{v}{a} = x^{\frac{1}{2}} J_1(2x^{\frac{1}{2}}) = \tilde{v}$$

from which the result will follow since $\tilde{v}'$ has zeros at $\lambda_n$. Equivalently, we prove

$$(4.80)\qquad \lim_{a\to 0} \frac{w}{a} = \frac{J_1(2x^{\frac{1}{2}})}{x^{\frac{1}{2}}} = \tilde{w}.$$

Defining

$$(4.81)\qquad y = w/a$$

and noting that $\tilde{w}$ is a solution of

$$(4.82)\qquad \tilde{w} = 1 - \int_0^x (1 - \xi/x)\,\tilde{w}\,d\xi$$

it follows that

$$(4.83)\qquad y - \tilde{w} = -\int_0^x (1 - \xi/x)\left(\frac{y}{(a^2y^2+1)^{\frac{1}{2}}} - \tilde{w}\right)d\xi$$

$$= -\int_0^x (1 - \xi/x)\left[(y - \tilde{w}) + \left(\frac{y}{(a^2y^2+1)^{\frac{1}{2}}} - y\right)\right]d\xi\,.$$

Since $|y| = |w/a| \leq 1$ (cf. (4.71)), and

$$(4.84)\qquad \left|\frac{y}{(a^2y^2+1)^{\frac{1}{2}}} - y\right| \leq a^2y^2 \leq a^2$$

equ. (4.83) can be replaced by the inequality

$$(4.85)\qquad |y - w| \leq \int_0^x (1 - \xi/x)(|y - \tilde{w}| + a^2)\,d\xi.$$

This inequality implies, in the usual way, that

$$(4.86)\qquad \|y - \tilde{w}\|_2 \leq \frac{a^2}{2e}\,.$$

Thus $z_n(a) \to \lambda_n$ as $a \to 0$. This completes the proof of part 2 of theorem (4.2).

In order to show that $z_n(a)$ is a differentiable function of $a$, it suffices to recall that by definition

$$(4.87)\qquad v'(z_n(a), a) \equiv 0.$$

If (4.87) is formally differentiated with respect to $a$, it is found that

$$(4.88)\qquad v''\frac{dz_n}{da} + \Delta' = 0.$$

Thus $z_n$ is a differentiable function of $a$ if $v''(z_n(a),a) \neq 0$. However, $v'' = -v/(v^2 + x^2)^{\frac{1}{2}} \neq 0$ since $v(z_n,a) = v'(z_n,a) = 0$ would imply that $v \equiv 0$.

Thus $\frac{dz_n}{da}$ can be written explicitly as

$$(4.89)\qquad \frac{dz_n}{da} = \frac{\Delta'(z_n(a),a)}{p(z_n(a),a)\,v(z_n(a),a)}, \quad p = v/(v^2 + x^2)^{\frac{1}{2}}.$$

In order to show that $dz_n/da \geq 0$ it suffices to show that both $\Delta'(z_n,a)$ and $v(z_n,a)$ have the same sign. In fact, it will be shown that $\Delta'$ and $v$ are related to each other as depicted in fig. 4.3, where $\Delta'(\beta_{i+1}) = \Delta'(\alpha_i) = 0$, $(i = 0,1,2,\ldots)$.

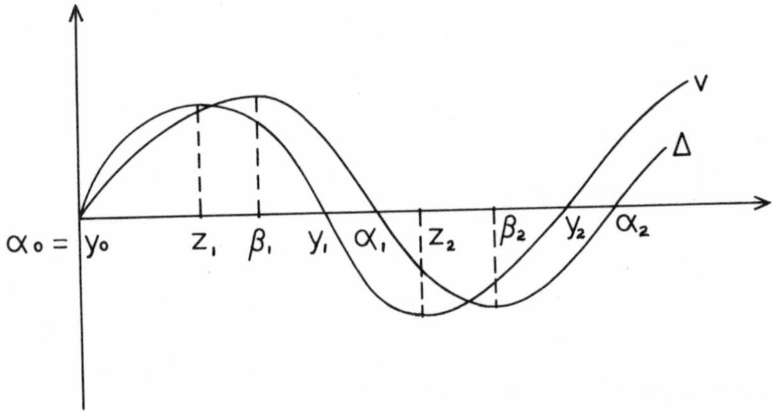


fig. 4.3

This result is actually sufficient to prove $dz_n/da \geq 0$ since it implies that $\Delta'(z_n)$ and $v(z_n)$ have the same sign.

In order to show that $v$ and $\Delta$ are related in this way, two lemmas are necessary.

Lemma (4.2): Between consecutive zeros of $\Delta$ there is at least one zero part of $v$.

Proof: Define

$$(4.90) \qquad p = \sqrt{(x^2 + v^2)}$$

so that equs. (4.30) and (4.52) can be written

$$(4.91) \qquad v'' + pv = 0$$

$$(4.92) \qquad \Delta'' + x^2 p^3 \Delta = 0.$$

Since

$$(4.93) \qquad p \geq x^2 p^3$$

the Sturm comparison theorem implies the result. Q.E.D.

Lemma (4.3): If $a > 0$, then $0 = y_0 = \alpha_0 < z_1 < \beta_1 < y_1 < \ldots < y_n < \alpha_n < z_{n+1} < \beta_{n+1} < y_{n+1} < \ldots$.

Proof: In order to prove this lemma, we require three identities. The first identity is obtained by multiplying equ. (4.92) by $v$ and multiplying equ. (4.91) by $\Delta$, subtracting the resulting expressions and integrating from $x_0$ to $x_1$. The resulting identity has the form

$$(4.94) \qquad (\Delta v' - v\Delta')\Big|_{x_0}^{x_1} + \int_{x_1}^{x_0} p^3 v^3 \Delta d\xi = 0.$$

In addition, $v'$ satisfies the equation

$$(4.95) \qquad (v')'' + x^2 p^3 v' - xp^3 v = 0$$

from which it follows that

$$(4.96) \qquad (\Delta v'' - v'\Delta')\Big|_{x_0}^{x_1} - \int_{x_1}^{x_0} \xi p^3 v \Delta d\xi = 0.$$

In order to obtain the last identity, define the function

$$(4.97) \qquad \tilde{v} = (x + \alpha_n) v' - 2v.$$

The function $\tilde{v}$ satisfies the equation

$$(4.98) \qquad \tilde{v}'' + x^2 p^3 \tilde{v} + x(x - \alpha_n) p^3 v = 0$$

from which it follows that

$$(4.99) \qquad (\Delta \tilde{v}' - \tilde{v}\Delta')\Big|_{x_0}^{x_1} + \int_{x_0}^{x_1} \xi(\xi - \alpha_n) p^3 v \Delta d\xi = 0.$$

Evaluating equ. (4.94) at $(\alpha_n, z_{n+1})$, equ. (4.96) at $(y_n, z_{n+1})$, and equ. (4.99) at $(y_n, y_{n+1})$, we find

$$(4.100) \qquad v(z_{n+1})\Delta'(z_{n+1}) = v(\alpha_n)\Delta'(\alpha_n) + \int_{\alpha_n}^{z_{n+1}} p^3 v^3 \Delta d\xi$$

$$(4.101) \qquad \Delta(z_{n+1}) v''(z_{n+1}) + \Delta'(y_n) v'(y_n) - \int_{y_n}^{z_{n+1}} \xi p^3 v \Delta d\xi = 0$$

$$(4.102) \qquad [\Delta(y_{n+1}) + (y_{n+1} + \alpha_n)\Delta'(y_{n+1})]\, v'(y_{n+1}) =$$

$$\int_{y_n}^{y_{n+1}} \xi(\xi - \alpha_n) p^3 v \Delta d\xi + [\Delta(y_n) + (y_n + \alpha_n)\Delta'(y_n)]\, v'(y_n).$$

We begin the proof of the lemma by showing that $0 = y_0 = \alpha_0 < z_1 < \beta_1 < y_1 < \alpha_1$, i.e., we shall show that fig. 4.3 is correct in the first interval. It follows immediately that $y_1 < \alpha_1$ from lemma (4.2). In addition, $\Delta'(z_1) > 0$ since equ. (4.100) implies that

$$(4.103) \qquad v(z_1)\Delta'(z_1) > 0,$$

and $v(z_1) > 0$. However, in view of the fact that $\Delta'$ has exactly one zero between $\alpha_0$ and $\alpha_1$, we conclude that $z_1 < \beta_1$. It remains to show that

$\beta_1 < y_1$. However, equ. (4.102) yields

$$(4.104) \qquad [\Delta(y_1) + y_1\Delta'(y_1)]\, v'(y_1) = \int_0^{y_1} \xi^2 p^3 v \Delta d\xi > 0$$

so that

$$(4.105) \qquad y_1 \Delta'(y_1) v'(y_1) > -\Delta(y_1) v'(y_1) > 0$$

since $v'(y_1) < 0$ and $\Delta(y_1) > 0$. The conclusion is that $\Delta'(y_1) < 0$. Thus the zero of $\Delta'$ must occur at some point to the left of $y_1$, i.e., $\beta_1 < y_1$. This completes the proof of $0 = y_0 = \alpha_0 < z_1 < \beta_1 < y_1 < \alpha_1$. However, it should be noted that the inequality

$$(4.106) \qquad [\Delta(y_1) + (y_1 + \alpha_1)\Delta'(y_1)]\, v'(y_1) > 0$$

also follows from (4.104) and the fact that $\alpha_1 > 0$. We now extend the preceding arguments to the second interval, i.e., we show that $\alpha_1 < z_2 < \beta_2 < y_2 < \alpha_2$. The fact that $y_2 < \alpha_2$ follows from lemma (4.2). The fact that $z_2 < \beta_2$ follows from the identity (4.100) if $\alpha_1 < y_2$, i.e.,

$$(4.107) \qquad v(z_2)\Delta'(z_2) = v(\alpha_1)\Delta'(\alpha_1) + \int_{\alpha_1}^{z_2} p^3 v^3 \Delta d\xi$$

where $v(\alpha_1) < 0$, $\Delta'(\alpha_1) < 0$ and $v < 0$, $\Delta < 0$ for $\alpha_1 < \xi < z_2$. Thus

$$(4.108) \qquad v(z_2)\Delta'(z_2) > 0.$$

However, since $v(z_2) < 0$ (4.108) shows that $\Delta'(z_2) < 0$ which implies that $\beta_2 > z_2$. Now we show that $\beta_2 < y_2$. This follows from (4.102), i.e.,

$$(4.109) \qquad [\Delta(y_2) + (y_2 + \alpha_1)\Delta'(y_2)]\, v'(y_2) = \int_{y_1}^{y_2} \xi(\xi - \alpha_1) p^3 v \Delta d\xi$$

$$+ [\Delta(y_1) + (y_1 + \alpha_1)\Delta'(y_1)]\, v'(y_1).$$

Note that (cf. (4.106)) the last term on the right of equ. (4.109) is positive. The positivity of the integral follows after breaking it up into an integral from $y_1$ to $\alpha_1$ and an integral from $\alpha_1$ to $y_2$. Thus

$$(4.110) \qquad (y_2 + \alpha_2)\Delta'(y_2)\, v'(y_2) > -\Delta(y_2)\, v'(y_2) > 0.$$

This inequality yields

$$(4.111) \qquad \Delta'(y_2)\, v'(y_2) > 0$$

or since $v'(y_2) > 0$ we have $\Delta'(y_2) > 0$. Thus $\beta_2 < y_2$.

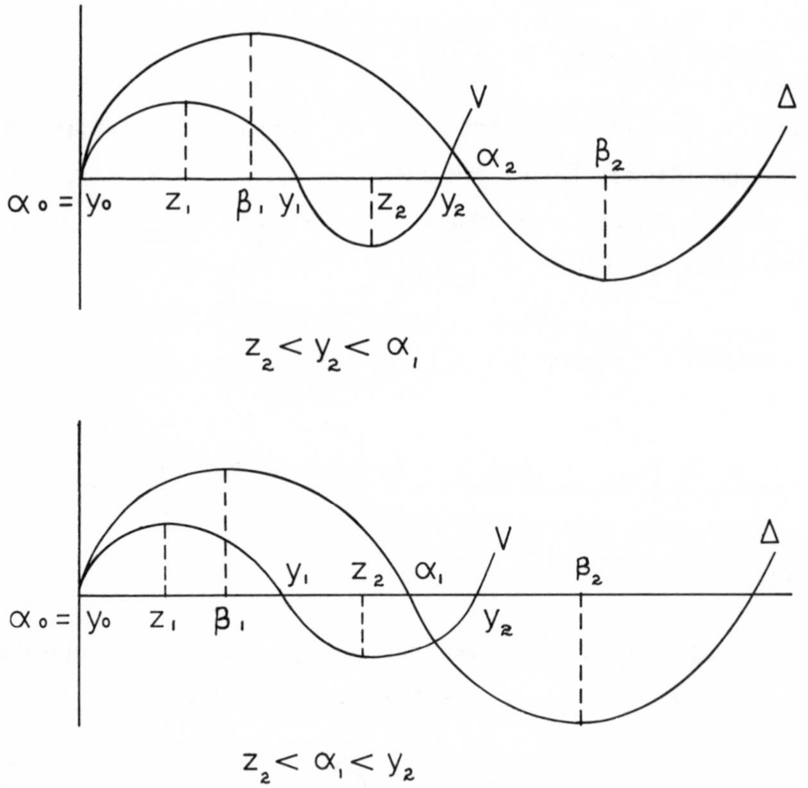


fig. 4.4

It is now necessary in order to complete the proof for the second interval to show that $\alpha_1 < z_2$. The argument relies on the identity (4.101). Assume that $z_2 \leq \alpha_1$ so that the curve appears as in fig. 4.4. It is a consequence of the assumption $z_2 \leq \alpha_1$ that (cf. fig. 4.4)

$$(4.112) \qquad \Delta(z_2)\, v''(z_2) = -p(z_2)\Delta(z_2)\, v(z_2) \geq 0.$$

However, the identity (4.101) yields

$$(4.113) \qquad \Delta(z_2)\, v''(z_2) + \Delta'(y_1)\, v'(y_1) - \int_{y_1}^{z_2} \xi p^3 v \Delta d\xi = 0$$

or equivalently $\Delta(z_2)\, v''(z_2) < 0$. This contradiction proves that $\alpha_1 < z_2$. This completes the proof for the second interval. In addition it follows from the fact that the right side of (4.109) is positive, $\alpha_2 > \alpha_1$, and (4.111) that

$$(4.114) \qquad [\Delta(y_2) + (y_2 + \alpha_2)\Delta'(y_2)]\, v'(y_2) > 0.$$

The argument proceeds by induction. Q.E.D.

This completes the proof of part 3 of theorem 4.2, and hence of the theorem itself. Q.E.D.

Reference


1 I.I. Kolodner, Heavy Rotating String - A Nonlinear Eigenvalue Problem. Comm. on Pure and Appl. Math. 8 (1955), 395-408.

# 5 Existence of positive solutions

In previous chapters emphasis has been on the existence or non-existence of multiple solutions rather than the behaviour of these solutions. It is clear, however, from Chapters 2 and 4 that a variety of different behaviours can be expected. Indeed the solutions discussed in Chapter 2 were all of one sign, while the discussion of the rotating string problem indicated the existence of solutions with multiple zeros. In addition, in Chapter 3 it was indicated that, at least under certain circumstances, the circular membrane problem led to solutions of one sign. In this chapter we discuss the question of the existence anduniqueness of positive solutions to problems of the form

$$(5.1) \qquad Lu = \lambda f(x,u)$$
$$u(0) = u(1) = 0$$

where

$$(5.2) \qquad L = -\frac{d}{dx} a_1(x) \frac{d}{dx} + a_0(x)$$

with $a_1(x)$ differentiable, $a_1(x) > 0$, and $a_0(x) \geq 0$ for $0 \leq x \leq 1$. The following lemma is an immediate consequence of the maximum principle for linear second order operators.

Lemma (5.1): If $L\emptyset > 0$ and $\emptyset(0) = \emptyset(1) = 0$, then $\emptyset(x) > 0$ in the interval $0 < x < 1$.

However, for the purposes of the following discussion, a somewhat stronger result is needed.

Theorem (Positivity Lemma) (5.1): Let $\rho(x) > 0$ for $0 < x < 1$ and assume that $\emptyset(x)$ is twice continuously differentiable in this interval.

If

$$(5.3)\qquad L\phi - \lambda\rho(x)\phi > 0$$
$$\phi(0) = \phi(1) = 0$$

then $\phi(x) > 0$ for $0 < x < 1$ iff $\lambda < \mu_1$ where $\mu_1$ is the smallest eigen-value of

$$(5.4)\qquad L\psi - \mu\rho(x)\psi = 0$$
$$\psi(0) = \psi(1) = 0.$$

Proof: By assumption $\phi$ satisfies

$$(5.5)\qquad L\phi - \lambda\rho(x)\phi = P(x)$$
$$\phi(0) = \phi(1) = 0$$

and $P(x) > 0$ for $0 < x < 1$. It is well known that

$$(5.6)\qquad \mu_1 = \min \frac{\int_0^1 [a_1(x)(g')^2 + a_0 g^2]dx}{\int_0^1 \rho(x) g^2 dx}$$

where the minimum is taken over all piecewise differentiable functions $g(x)$ satisfying $g(0) = g(1) = 0$. Thus for any particular choice of $g$ (5.6) implies that

$$(5.7)\qquad \frac{\int_0^1 [a_1(x)(g')^2 + a_0 g^2]dx}{\int_0^1 \rho(x) g^2 dx} \geq \mu_1$$

or assuming $\lambda < \mu_1$

$$(5.8)\qquad \int_0^1 [a_1(x)(g')^2 + (a_0(x) - \lambda\rho(x)g^2]dx > 0.$$

The function $\phi$ minimizes

$$(5.9)\qquad \int_0^1 [a_1(g')^2 + (a_0 - \lambda\rho)g^2 - 2P(x)g]\,dx.$$

Assume $\phi$ is not positive. Then $|\phi|$ is a piecewise differentiable function and we note that

$$(5.10)\qquad \int_0^1 [a_1(|\phi|')^2 + (a_0 - \lambda\rho)|\phi|^2]\,dx = \int_0^1 [a_1(\phi')^2 + (a_0 - \lambda\rho)\phi^2]\,dx.$$

However,

$$(5.11)\qquad \int_0^1 P(x)\phi\,dx < \int_0^1 P(x)|\phi(x)|\,dx$$

which implies that

$$(5.12)\qquad \int_0^1 [a_1(\phi')^2 + (a_0 - \lambda\rho)\phi^2 - 2P\phi]\,dx$$

$$> \int_0^1 [a_1(|\phi|')^2 + (a_0 - \lambda\rho)|\phi|^2 - 2P|\phi|]\,dx.$$

Thus $\phi$ could not be the minimum. This contradiction shows that $\phi(x) \geq 0$. The function $\phi(x) > 0$ for $0 < x < 1$ since if $\phi = 0$ for some value of $x$ (5.5) would imply that $\phi'' < 0$ at a minimum. Q.E.D.

This theorem can be replaced by a weaker result.

Theorem (5.2): Let $\rho(x) > 0$ for $0 < x < 1$ and assume that $\phi(x)$ is twice continuously differentiable in this interval. If

$$(5.13)\qquad \begin{aligned} &L\phi - \lambda\rho\phi \geq 0 \\ &\phi(0) = \phi(1) = 0 \end{aligned}$$

then $\phi(x) \geq 0$ for $0 < x < 1$ iff $\lambda < \mu_1$ where $\mu_1$ is the smallest eigenvalue of (5.4).

The preceding positivity lemma is the key result in deciding when problem (5.1) has positive solutions. For simplicity we will denote by $\Lambda$ the set of $\lambda$ for which (5.1) has a positive solution and let

(5.14) $\lambda^* = \operatorname{lub}_{\lambda \in \Lambda} \lambda .$

At various points in the following discussion it will be assumed that $f(x,u)$ satisfies the conditions

(1) $f(x,\emptyset)$ is continuous when $0 < x < 1$ and $\emptyset \geq 0$.

(2) $f(x,0) = f_0(x) > 0$.

(3) $f(x,\emptyset) > f(x,\psi)$ if $\emptyset > \psi \geq 0$.

Theorem (5.3): If $f(x,\emptyset) > 0$ when $0 < x < 1$ and $\emptyset > 0$ then $\lambda \epsilon \Lambda$ implies that $\lambda > 0$.

Proof: The proof is by contradiction. Assume $u(x) > 0$ is a solution of (5.1) with $\lambda < 0$. It follows that

(5.15) $$L(-u) = -\lambda f(x,u) > 0$$
$$-u(0) = -u(1) = 0.$$

However, the positivity lemma implies that $-u > 0$. This contradiction proves the theorem. Q.E.D.

We are now in a position to determine a necessary and sufficient condition for (5.1) to have a positive solution. Define the sequence $\{u_n\}$ where

(5.16) $$Lu_n = \lambda f(x,u_{n-1})$$
$$u_n(0) = u_n(1) = 0$$

and $u_0 \equiv 0$.

Definition: $\underline{u}(x,\lambda)$ is a minimal positive solution if $\underline{u}(x,\lambda) > 0$ for $0 < x < 1$ and $\underline{u}(x,\lambda) \leq u(x,\lambda)$ for any positive solution of (5.1).

Theorem (5.4): If $f$ satisfies the conditions (1), (2) and (3), then $\lambda > 0$ is in $\Lambda$ iff the sequence $\{u_n(x,\lambda)\}$ is uniformly bounded independent of $n$. If $\lambda \in \Lambda$ then $\{u_n(x,\lambda)\}$ converges uniformly to $\underline{u}(x,\lambda)$.

Proof: The fact that the function $u_1$ is positive follows from

$$Lu_1 = \lambda f(x,0) = \lambda f_0(x) > 0$$

and the positivity lemma. Moreover, the fact that

$$L(u_2 - u_1) = \lambda(f(x,u_1) - f(x,0)) > 0$$

indicates that $u_2 > u_1$. By induction it is shown that the sequence $\{u_n\}$ is monotone increasing. If the functions $\{u_n\}$ are bounded, they converge to some function $\underline{u}(x,\lambda) \leq M$. In addition, $\underline{u}(x,\lambda)$ is continuous since not only the $u_n$ are bounded, but the differential equation guarantees that the derivatives are bounded. Rewriting the differential equation as a nonlinear integral equation

$$(5.17) \qquad u_n(x) = \lambda \int_0^1 G(x,\xi) f(\xi u_n) d\xi$$

$(G(x,\xi) \geq 0)$ and noting that

$$\int_0^1 G(x,\xi) f(\xi,u_n) d\xi \leq \int_0^1 G(x,\xi) f(\xi,M) d\xi,$$

the Dominated Convergence theorem implies that

$$(5.18) \qquad \underline{u}(x,\lambda) = \lambda \int_0^1 G(x,\xi) f(\xi,\underline{u}) d\xi.$$

After differentiating (5.18) we find that $\underline{u}(x,\lambda)$ is a positive function satisfying (5.1). Thus $\lambda \in \Lambda$.

In order to prove the necessity of the theorem, assume that $\lambda > 0$ and $\lambda \in \Lambda$ and assume that $u(x,\lambda)$ is a corresponding positive solution. The function $u(x,\lambda) > 0$. In addition the fact that

$$L(u - u_1) = \lambda(f(x,u) - f(x,0)) > 0$$

implies that $u > u_1$. By induction it is easily shown that $u > u_n$, i.e., $\{u_n\}$ is bounded. Note that it also follows that $u \geq \underline{u}$. Thus $\underline{u}$ is the minimal positive solution. Q.E.D.

The following set of theorems are devoted to finding conditions on $f$ so that (5.1) has a positive solution.

Theorem (5.5): Assume $f$ satisfies the conditions (1), (2) and (3) and assume $F(x, u)$ is any other function with the property that $F(x,\emptyset) > f(x,\psi)$ if $\emptyset > \psi \geq 0$. If

$$(5.19) \qquad \begin{aligned} Lv &= \lambda F(x,v) \\ v(0) &= v(1) = 0 \end{aligned}$$

has a positive solution for some fixed $\lambda > 0$, then $\lambda \in \Lambda$ and the minimal positive solution of (5.1) satisfies $\underline{u}(x,\lambda) \leq v(x,\lambda)$.

Proof: In view of theorem (5.4), it suffices to show that the sequence $\{u_n\}$ is uniformly bounded. However, $v \geq u_0 = 0$ and

$$L(v - u_1) = \lambda(F(x,v) - f(x,0)) > 0$$

so that $v > u_1$. Induction shows that $v > u_n$. Q.E.D.

Theorem(5.6): Assume $f$ satisfies the conditions (1), (2) and (3) and assume that $\lambda' \in \Lambda$ and $\lambda' > 0$. The interval $0 < \lambda \leq \lambda'$ is in $\Lambda$. Moreover, $\underline{u}$ is an increasing function of $\lambda$.

Proof: For any fixed value $\lambda$ in the interval $0 < \lambda < \lambda'$, define

$$F(x,\emptyset) = \frac{\lambda'}{\lambda} f(x,\emptyset).$$

The function $F$ has the property that $F(x,\emptyset) > f(x,\psi)$ if $\emptyset > \psi \geq 0$. Moreover,

$$Lv = \lambda F(x,v) = \lambda' f(x,v)$$
$$v(0) = v(1) = 0$$

has a positive solution. By theorem (5.5) equ. (5.1) has a positive solution. In addition, theorem (5.5) indicates that $\underline{u}(x,\lambda) = \underline{u}(x,\lambda') = \underline{v}(x,\lambda)$. Q.E.D.

Theorem (5.7): Assume $f$ satisfies conditions (1), (2) and (3) and assume there exists a function $F(x) > 0$ such that $f(x,\emptyset) < F(x)$ for all $\emptyset > 0$. Then all $\lambda > 0$ are in $\Lambda$, (i.e., $\lambda^* = \infty$).

Proof: The problem

$$Lv = \lambda F(x)$$
$$v(0) = v(1) = 0$$

has a positive solution for any $\lambda > 0$ since

$$v = \lambda \int_0^1 G(x,\xi) F(\xi) d\xi > 0.$$

By theorem (5.6) all $\lambda > 0$ are in $\Lambda$. Q.E.D.

Theorem (5.8): Assume $f$ satisfies conditions (1), (2) and (3) and assume there exists functions $F(x) > 0$ and $\rho(x) > 0$ such that

$$f(x,\emptyset) < F(x) + \rho(x)\emptyset$$

for all $\emptyset > 0$. Then $\Lambda$ contains all $\lambda$ such that $0 < \lambda < \mu_1(\rho)$ where $\mu_1$ is the smallest eigenvalue of

$$Lv - \mu \rho v = 0$$
$$v(0) = v(1) = 0$$

$(\mu_1(\rho) \leq \lambda^*)$.

Proof: By the positivity lemma

$$Lv - \mu \rho v = \mu F(x)$$
$$v(0) = v(1) = 0$$

has a positive solution when $\mu < \mu_1$. Thus theorem (5.5) implies that (5.1) has a positive solution when $0 < \lambda < \mu_1(\rho)$. Q.E.D.

Theorem (5.9): Assume $f$ satisfies conditions (1), (2) and (3). Let $\lambda^* < \infty$ be a least upper bound on $\Lambda$. If $F(x,\emptyset) > f(x,\psi)$ when $\emptyset > \psi \geq 0$ then

$$Lv = \lambda F(x,v)$$
$$v(0) = v(1) = 0$$

has no positive solution for $\lambda > \lambda^*$. In particular, if $f(x,\emptyset) > F(x) + \rho(x)\emptyset$ $(F > 0, \ \rho > 0)$ when $\emptyset > 0$ then $\lambda^* \leq \mu_1(\rho)$.

Proof: If

$$Lv = \lambda F(x,v)$$
$$v(0) = v(1) = 0$$

has a solution for $\lambda > \lambda^*$, theorem (5.6) implies that (5.1) has a solution for $\lambda > \lambda^*$. This is a contradiction since $\lambda^*$ is an upper bound of $\Lambda$. In order to prove the second part of the lemma, simply note that the positivity lemma implies that

$$Lv = \lambda(F(x) + \rho v)$$
$$v(0) = v(1) = 0$$

has no positive solution when $\lambda \geq \mu_1$. Thus by the first part of the lemma (5.1) has no positive solution for $\lambda \geq \mu_1$. Q.E.D.

Theorem (5.10): Assume $f$ satisfies conditions (1), (2) and (3) and assume that $f(x,\emptyset) < \rho(x)\emptyset$ for all $\emptyset > 0$. Then there exists no positive solution of (5.1) for $0 < \lambda < \mu_1$.

Proof: Assume there is a positive solution. Then

$$L(-u) - \lambda\rho(-u) = \lambda(\rho u - f(x,u)) > 0$$

so that the positivity lemma implies $-u > 0$. This contradiction proves the theorem. Q.E.D.

The preceding theorems give a variety of tests for determining whether a problem of the form (5.1) has a positive solution. We now consider the question of multiplicity of positive solutions. For this purpose it is convenient to assume a stronger monotonicity condition than that furnished by condition (3). Thus it will be assumed that

$$(3') \qquad \frac{\partial f}{\partial \phi} > 0 \text{ and continuous if } \phi > 0.$$

Note that condition (3') implies condition (3). Our uniqueness result rests on the following theorem.

Theorem (5.11): Assume $f$ satisfies the conditions (1), (2) and (3') and assume (5.1) has positive solutions in the interval $0 < \lambda < \lambda^*$. Each $\lambda$ in the interval satisfies the condition $\lambda < \mu_1$ where $\mu_1$ is the smallest eigenvalue of

$$(5.20) \qquad L\psi - \mu f_u(x,\underline{u}(x,\lambda))\psi = 0$$
$$\psi(0) = \psi(1) = 0.$$

Proof: Assume $\lambda$ and $\lambda + \Delta\lambda$ are in the interval $0 < \lambda < \lambda^*$. Define the function

$$(5.21) \qquad V_n(x,\lambda,\Delta\lambda) = \frac{u_n(x,\lambda + \Delta\lambda) - u_n(x,\lambda)}{\Delta\lambda}$$

so that $V_n$ satisfies the equation

$$(5.22) \qquad LV_n = \frac{(\lambda+\Delta\lambda)\, f(x,u_{n-1}(x,\lambda+\Delta\lambda)) - \lambda f(x,u_{n-1}(x,\lambda))}{\Delta\lambda}$$

the mean value theorem implies that $V_n$ satisfies

$$(5.23) \qquad LV_n = \lambda f_u(x, (1-\theta)u_{n-1}(x,\lambda) - \theta u_{n-1}(x,\lambda + \Delta\lambda))V_{n-1}$$
$$+ f(x,u_{n-1}(x,\lambda + \Delta\lambda))$$

$$V_n(0) = V_n(1) = 0$$

where $0 < \theta < 1$. It may be shown by standard techniques that

i) $\lim_{\Delta\lambda \to 0} V_n$ exists,

ii) $V_n$ is a continuous function of $\lambda$, and

iii) $V_n(x,\lambda,\Delta\lambda) \to \underline{V}(x,\lambda) = \dfrac{\partial \underline{u}(x,\lambda)}{\partial\lambda}$

as $\lambda \to 0$ where $\underline{V}(x,\lambda)$ is continuous and satisfies

$$(5.24) \qquad L\underline{V} = \lambda f_u(x,\underline{u}(x,\lambda))\,\underline{V}(x,\lambda) + f(x,\underline{u}(x,\lambda))$$
$$\underline{V}(0) = V(1) = 0.$$

Since $\underline{u}$ is an increasing function of $\lambda$ (theorem (5.6)), it follows that $\underline{V} = \partial\underline{u}/\partial\lambda \geq 0$. Actually $\underline{V} > 0$ for $0 < x < 1$ since if $\underline{V} = 0$ equ.(5.24) would imply that $\underline{V}'' < 0$ at a minimum. Since $f_u > 0$ and $f > 0$ the positivity lemma implies that (5.24) has a positive solution only in the case $\lambda < \mu_1$. Q.E.D.

The preceding theorem enables us to prove one result on the uniqueness of positive solutions.

Definition: $f(x,u)$ is concave if $f_u(x,\emptyset) < f_u(x,\psi)$ when $\emptyset > \psi \geq 0$.

Theorem (5.12): Assume $f$ satisfies the conditions (1), (2) and (3') and assume that $f$ is concave. Then (5.1) has at most one positive solution.

Proof: Let $\underline{u}(x,\lambda)$ be the minimal positive solution and assume $u(x,\lambda)$ is some other positive solution, i.e., $u(x,\lambda) - \underline{u}(x,\lambda) \geq 0$ for $0 < x < 1$. The function $u - \underline{u}$ satisfies the equation

$$L(u - \underline{u}) = \lambda(f(x,u) - f(x,\underline{u})) = \lambda f_u(x,\underline{u} + \theta(u - \underline{u}))(u - \underline{u})$$

where $0 < \theta < 1$. The concavity of $f$ implies that

$$L(\underline{u} - u) \geq \lambda f_u(x,\underline{u})(\underline{u} - u).$$

Theorem (5.11) implies that $\lambda < \mu_1$ where $\mu_1$ is the smallest eigenvalue of

$$L\psi - \mu f_u(x,\underline{u})\psi = 0$$
$$\psi(0) = \psi(1) = 0.$$

The weak positivity lemma (theorem (5.2)) implies that $\underline{u} > u$. Thus $u \equiv \underline{u}$.
Q.E.D.

The above theorem requires a stronger condition on f than the strict monotonicity condition (3'). In fact, it cannot be expected that in general there is only one positive solution. The type of approach discussed in this chapter can be used to prove the existence of a 'minimal comparison' solution for the circular membrane problem discussed in chapter 3.

## References


1 H.B. Keller and D.S. Cohen, Some Positone Problems Suggested By Nonlinear Heat Generation, J.Math. and Mechs. 16 (1967), 1361-1376.

2 H.B. Keller, Positive Solutions of Some Nonlinear Eigenvalue Problems, J. Math. and Mechs. 19 (1969), 279-295.

# 6 Bifurcation theory for second order ordinary differential equations. Application to the inextensible elastica

In this chapter we develop criteria to determine the bifurcation points and the nature of the branches close to these bifurcation points for the problem

(6.1a) $\quad (I(t)u_t)_t + f(u,t,\lambda) = 0$

(6.1b) $\quad B_1(u) = \alpha_1 u_t(t_1) + \beta_1 u(t_1) = 0$

(6.1c) $\quad B_2(u) = \alpha_2 u_t(t_2) + \beta_2 u(t_2) = 0$

where $\alpha_1 \neq 0$ and $\alpha_2^2 + \beta_2^2 \neq 0$. Thus we assume the problem (6.1) has a solution $u_0(t,\lambda)$ for $t_1 \leq t \leq t_2$ and $\lambda$ in some interval $|\lambda - \lambda_0| \leq c$, and consider the question of whether there is another solution $u(t,\lambda)$ such that $u(t,\lambda) \to u_0(t,\lambda_0)$ as $\lambda \to \lambda_0$, but $u(t,\lambda) \not\equiv u_0(t,\lambda_0)$ for $0 < |\lambda - \lambda_0| < \epsilon$. In the problem treated in previous chapters, we have obtained global results. In the case of (6.1) however, the object is strictly local results, i.e., we want to decide when $\lambda_0$ is a bifurcation point, what the bifurcation curve looks like in the neighbourhood of the bifurcation point, and the behaviour of the solution when $\lambda$ is close to the bifurcation point. At the end of this chapter these results will be extended to obtain global results for the inextensible elastica.

In conjunction with the boundary value problem (6.1) it is convenient to consider the initial value problem (6.1a), (6.1b) and the additional condition

(6.2) $\quad u(t_1,\lambda) = a$

where we assume that $a$ is in some neighbourhood of $a_0$ where

(6.3) $\quad u_0(t,\lambda_0) = a_0.$

The existence of a unique solution to the initial value problem (6.1a), (6.1b) and (6.2) in the interval $t_1 \leq t \leq t_2$ follows from the standard existence and uniqueness theorem for ordinary differential equations. In order to simplify the notation define

$$\| \cdot \| = \max_{t_1 \leq t \leq t_2} | \cdot |.$$

Theorem (6.1): Assume $I(t) > 0$ and is differentiable for $t_1 \leq t \leq t_2$. Let $u_0(t,\lambda_0)$ be a solution of (6.1) for $t_1 \leq t < t_2$ and assume $f(u,t,\lambda)$ and $f_u(u,t,\lambda)$ are defined and continuous for $t_1 \leq t \leq t_2$, $\|u - u_0\| \leq \delta$, and $|\lambda - \lambda_0| \leq c$. Assume $u_0(t,\lambda_0) = a_0$. There exist positive constants, $\alpha$ and $\beta$, such that for every $a$ and $\lambda$ satisfying $|a - a_0| \leq \alpha$ and $|\lambda - \lambda_0| \leq \beta \leq c$ there exists a unique solution $u(t,\lambda,a)$ defined for $t_1 \leq t \leq t_2$ and satisfying the initial condition $u(t,\lambda,a) = a$ and $B_1(u(t,\lambda,a))=0$.

Assume $u(t,\lambda,a)$ is a solution of the initial value problem. The function $u(t,\lambda,a)$ will be a solution of the boundary value problem (6.1) if

$$(6.4) \qquad B_2 u(t,\lambda,a) = b(a,\lambda) = 0.$$

Conversely, if $b(a,\lambda) = 0$ then $u(t,\lambda,a)$ is a solution of (6.1).

Definition: The equation $b(a,\lambda) = 0$ is the bifurcation equation.

In view of the above discussion, we have the theorem

Theorem (6.2): The solution of the initial value problem is a solution of the boundary value problem iff $a$ and $\lambda$ are solutions of the bifurcation equation.

In view of our assumption that $u_0(t,\lambda)$ is a solution of (6.1), the bifurcation equation certainly has one solution in a neighbourhood of $\lambda_0$. In fact, if $b(a,\lambda)$ is a differentiable function of $a$ at $a = a_0$, $\lambda = \lambda_0$ and $b_a(a,\lambda) \neq 0$ at $(a_0,\lambda_0)$, then it is well known that the solution of $b(a,\lambda) = 0$, i.e., $a = a(\lambda)$, is unique. Thus bifurcation does not occur at $\lambda = \lambda_0$ in this case (cf. fig. 6.1). On the other hand, if $b_a(a_0,\lambda_0) = 0$ then the possibility exists that the solution of the bifurcation equation in a neighbourhood of $\lambda = \lambda_0$ is not unique, i.e., bifurcation occurs.

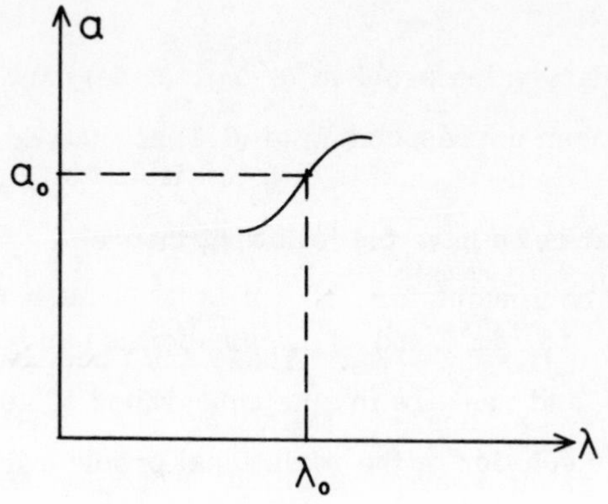


fig. 6.1

Theorem (6.3): Assume $b(a,\lambda)$ and $b_a(a,\lambda)$ are defined and continuous in a neighbourhood of $(a_0,\lambda_0)$. A necessary condition for bifurcation to occur at $\lambda = \lambda_0$ is that $b_a(a_0,\lambda_0) = 0$.

In order to actually evaluate $b_a(a,\lambda)$ for any value of $(a,\lambda)$ it is necessary to know the derivative of $u(t,\lambda,a)$ with respect to $a$. Consider the equ. (6.1a) with initial conditions $u(t_1,\lambda_0,a_0+\epsilon c) = a_0 + \epsilon c$, $B_1 u(t,\lambda_0,a_0+\epsilon c) = 0$. Under sufficiently strong conditions on $f(u,t,\lambda_0)$ the solution $u(t,\lambda_0,a_0+\epsilon c)$ will be differentiable with respect to $a$, i.e., with respect to $\epsilon$, and will satisfy

(6.5a) $\quad (I(t)\dot{u}_t)_t + f_u(u(t,\lambda_0,a_0),t,\lambda_0)\dot{u} = 0$

(6.5b) $\quad \dot{u}(t_1,\lambda_0,a_0) = c$

(6.5c) $\quad B_1\dot{u}(t_1,\lambda_0,a_0) = 0$

where $u = du(t,\lambda,a_0+\epsilon c)/d\epsilon$ evaluated at $\epsilon = 0$. Moreover, at $t = t_1$ we have $B_2\dot{u}(t_1,\lambda_0,a_0) = B_2 u_a(t,\lambda_0,a_0)c = b_a(a_0,\lambda_0)c$. If $b_a(a_0,\lambda_0) = 0$ it follows that (6.5a), (6.5c) and

(6.5d) $\quad B_2\dot{u}(t,\lambda_0,a_0) = 0$

has a non-trivial solution. Conversely, if $b_a(a_0,\lambda_0) \neq 0$ for any non-trivial initial data (6.5b) the boundary value problem has only the trivial solution.

Definition: The boundary value problem (6.5a), (6.5c), and (6.5d) is the variational problem corresponding to (6.1) at $(a_0,\lambda_0)$.

In view of the above remarks we have the following theorem.

Theorem (6.4): Assume $f$, $f_u$, and $f_{uu}$ are defined and continuous for $t_1 \leq t \leq t_2$ and $u,\lambda$, and $a$ are in a neighbourhood of $u_0,\lambda_0$, and $a_0$. The trivial solution is the only solution of the variational problem iff $b_a(a_0,\lambda_0) = 0$.

Definition: The degree of degeneracy of problem (6.1) is the number of linearly independent solutions of the variational problem.

The above results indicate that bifurcation is not possible unless (6.1) is degenerate, i.e., unless the corresponding variational problem has at least one non-trivial solution.

We now turn to the question of finding sufficient conditions for bifurcation to occur. Assume that (6.1) is degenerate of order 1 at $(a_0,\lambda_0)$, i.e., we assume the variational problem (6.5a), (6.5c), and (6.5d) has one non-trivial solution $\dot{u}(t,\lambda_0,a_0) = \emptyset(t)$ where $\emptyset(t)$ is made unique by the normalizing condition

(6.6) $\quad \emptyset(t_1) = 1.$

(This condition is possible since we have assumed that $\alpha_1 \neq 0$. Thus if $\emptyset(t_1) = 0$ it would follow from (6.5c) that $\emptyset_t(t_1) = 0$ which would imply that $\emptyset(t) \equiv 0$.) Define a new dependent variable

(6.7) $\quad u(t,\lambda) = u_0(t,\lambda) + a\,v(t,\lambda,a)$

where $u(t,\lambda)$ is a solution of (6.1a) and $u_0(t,\lambda)$ is a solution of the boundary value problem (6.1). The parameter $a$ will be described below. Since $u$ and $u_0$ are solutions of (6.1a), it follows that $v(t,\lambda,a)$ is a solution of

$$(6.8a) \qquad (Iv_t)_t + \frac{1}{a}[f(u_0(t,\lambda) + av(t,\lambda,a),t,\lambda) - f(u_0(t,\lambda),t,\lambda)] = 0.$$

For each $a \neq 0$ the existence and uniqueness theorem for the initial value problem implies the existence of a unique solution of (6.8a) satisfying the conditions

$$(6.8b) \qquad v(t_1,\lambda,a) = 1$$

$$(6.8c) \qquad B_1 v(t,\lambda,a) = 0.$$

If $u$, in addition to $u_0$, is to be a solution of the boundary value problem (6.1), it follows that $v(t,\lambda,a)$ must satisfy, in addition to (6.8b) and (6.8c), the equation

$$(6.8d) \qquad B_2 v(t,\lambda,a) = 0.$$

Thus far in our discussion we have assumed that $a \neq 0$. However, it is possible also to consider the point $a = 0$ since the equation (6.8a) can be defined so that it is continuous at $a = 0$. In fact, L'Hospital's rule implies that

$$\lim_{a \to 0} \frac{f(u_0 + av,t,\lambda) - f(u_0,t,\lambda)}{a} = f_u(u_0,t,\lambda)v$$

so that at $a = 0$ equ. (6.8a) becomes

$$(6.9) \qquad (Iv_t)_t + f_u(u_0(t,\lambda),t,\lambda)v = 0.$$

We claim that at $a = 0$, $\lambda = \lambda_0$ the problem (6.8a), (6.8b) and (6.8c) has a unique non-trivial solution satisfying (6.8d). This is an immediate result since at $a = 0$, $\lambda = \lambda_0$ the problem (6.9) with boundary conditions (6.8c) and (6.8d) is simply the variational problem and has a non-trivial solution by assumption. In fact, since the problem is degenerate of order 1, it follows that the unique solution of (6.8c), (6.8d) and (6.9) satisfying (6.8b) is just

$$(6.10) \qquad v(t,\lambda,a) \equiv \emptyset(t).$$

Equ. (6.8d) is just the bifurcation equation, and we have verified that it has a solution at $a = 0$, $\lambda = \lambda_0$. We wish to show that in every sufficiently small neighbourhood of $a = 0$, $\lambda = \lambda_0$ there is a solution of (6.8d) $a = a(\lambda) \not\equiv 0$. This will complete the proof that $\lambda_0$ is a bifurcation point since $u(t,\lambda) = u_0(t,\lambda) + a\,v(t,\lambda,a)$ is a solution $u(t,\lambda) \not\equiv u_0(t,\lambda)$ if $\lambda \neq \lambda_0$, and since $a(\lambda) \to 0$ as $\lambda \to 0$, we find that $u(t,\lambda) \to u_0(t,\lambda)$ as $\lambda \to \lambda_0$.

We can solve the bifurcation equation (6.8d) in the neighbourhood of $a = 0$, $\lambda = \lambda_0$, if $(B_2 v(t,\lambda,a))_a \neq 0$, i.e., if

$$(6.11) \qquad \alpha_2 v_{at}(t_2,\lambda_0,0) + \beta_2 v_a(t_2,\lambda_0,0) \neq 0.$$

The function $v_a$ satisfies

$$(6.12a) \qquad (I v_{at})_t + f_u(u_0(t,\lambda_0),t,\lambda_0)\, v_a = -\frac{1}{2} f_{uu}(u_0(t,\lambda_0),t,\lambda_0)\, \emptyset(t)^2$$

at $a = 0$, $\lambda = \lambda_0$ (where the right side of (6.12) has been evaluated by L'Hospital's rule). The function $v_a$ also satisfies the initial conditions

$$(6.12b) \qquad v_a(t_1,\lambda_0,0) = 0$$

$$(6.12c) \qquad B_1 v_a(t,\lambda_0,0) = 0.$$

This problem has a unique solution and with it we can evaluate $B_2 v_a(t,\lambda_0,0)$. If $B_2 v_a(t,\lambda_0,0) = 0$ then $v_a$ satisfies the same boundary conditions as $\emptyset(t)$. Note also that equ. (6.12a) is the inhomogeneous variational equation. Thus equ. (6.12a) has no solution satisfying (6.12c) and

$$(6.12d) \qquad B_2 v_a(t,\lambda_0,0) = b_a(0,\lambda_0) = 0$$

unless the right side is orthogonal to $\emptyset(t)$ (Fredholm alternative theorem), i.e., there is no solution unless

$$(6.13) \qquad \int_{t_1}^{t_2} f_{uu}(u_0(t,\lambda_0),t,\lambda_0)\, \emptyset^3(t)\, dt = 0.$$

If the condition (6.13) is violated, then every solution of (6.12a), (6.12b) and (6.12c) satisfies the condition $b_a(0,\lambda_0) \neq 0$. In this case there exists a unique solution $a = a(\lambda)$ in a neighbourhood of $a = 0$, $\lambda = \lambda_0$. The function $a(\lambda)$ is continuous and $a(\lambda_0) = 0$ (Implicit Function Theorem). The corresponding function (6.7) is a solution of the boundary value problem (6.1) for $\lambda$ near $\lambda_0$. It is continuous in $\lambda$, $u(t,\lambda_0) = u_0(t,\lambda_0)$, and $u(t,\lambda) \not\equiv u_0(t,\lambda)$ for $\lambda \neq \lambda_0$ since $v(t,\lambda,a) \not\equiv 0$ $(v(t,\lambda,a) = 1)$. This is, of course, all under the assumption that $a(\lambda) \not\equiv 0$ in a neighbourhood of $\lambda_0$.

It remains to decide whether or not $a(\lambda) = 0$. If $a(\lambda)$ is differentiable at $\lambda = \lambda_0$, then

$$(6.14) \qquad a(\lambda) = a'(\lambda_0)(\lambda - \lambda_0) + 0(\lambda - \lambda_0).$$

Moreover, it follows that since $b(a(\lambda),\lambda) = 0$ for $\lambda$ in a neighbourhood of $\lambda_0$ that

$$(6.15) \qquad b_a(0,\lambda_0)\, a'(\lambda_0) + b_\lambda(0,\lambda_0) = 0$$

if $b(0,\lambda)$ is differentiable with respect to $\lambda$. Thus it follows from (6.14) that

$$(6.16) \qquad a(\lambda) = -\frac{b_\lambda(0,\lambda_0)}{b_a(0,\lambda_0)}(\lambda - \lambda_0) + 0(\lambda - \lambda_0).$$

In order to determine $b_\lambda(0,\lambda_0)$, we assume that $u_{0\lambda}$ and $f_{u\lambda}$ exist and are continuous. In this case we find that $v_\lambda(t,\lambda_0,0)$ is a solution of

$$(6.17a) \qquad (Iv_{\lambda t})_t + f_u(u_0(t,\lambda_0),t,\lambda_0)\, v_\lambda = -f_{uu}(u_0(t,\lambda_0),t,\lambda_0)\, u_{0\lambda}$$

$$- f_{u\lambda}(u_0(t,\lambda_0),t,\lambda_0)\, \phi(t)$$

$$(6.17b) \qquad v_\lambda(t_1,\lambda_0,0) = 0$$

$$(6.17c) \qquad B_1 v_\lambda(t,\lambda_0,0) = 0.$$

Just as above

$$(6.17d) \qquad B_2 v_\lambda(t,\lambda_0,0) = b_\lambda(0,\lambda_0) = 0.$$

This can happen only in the case where the right side of (6.17a) is orthogonal to $\phi(t)$, i.e., only in the case

$$(6.18) \quad \int_{t_1}^{t_2} [f_{uu}(u_0(t,\lambda_0),t,\lambda_0)\, u_{0\lambda}(t,\lambda_0) + f_{u\lambda}(u_0(t,\lambda_0),t,\lambda_0)\, \emptyset(t)]\, \emptyset(t)\, dt = 0.$$

Conversely, if (6.18) does not vanish, then $b_\lambda(0,\lambda_0) \neq 0$ and bifurcation occurs. These results can be summarised in the following theorem.

Theorem (6.5): Assume $f$, $f_u$, $f_{uu}$ and $f_{u\lambda}$ exist and are continuous for $t_1 \leq t \leq t_2$ and $u,\lambda$ in a neighbourhood of $u_0,\lambda_0$, where $u_0(t,\lambda)$ is a solution of (6.1) for $\lambda$ in a neighbourhood of $\lambda_0$. In addition, assume $u_{0\lambda}(t,\lambda)$ exists and is continuous for $t_1 \leq t \leq t_2$.

If the variational problem at $u_0(t,\lambda_0)$ has a unique nontrivial solution $\emptyset(t)$ with $\emptyset(t_1) = 0'$ and

$$(6.19) \quad \int_{t_1}^{t_2} f_{uu}(u_0(t,\lambda_0),t,\lambda_0)\, \emptyset^3(t)\, dt \neq 0$$

$$(6.20) \quad \int_{t_1}^{t_2} [f_{uu}(u_0(t,\lambda_0),t,\lambda_0)\, u_{0\lambda}(t,\lambda_0) + f_{u\lambda}(u_0(t,\lambda_0),t,\lambda_0)\, \emptyset(t)]\, \emptyset(t)\, dt \neq 0$$

then bifurcation occurs at $\lambda = \lambda_0$. Moreover, there is exactly one branch of solutions bifurcating from $\lambda_0$.

The situation described above corresponds to fig. 6.2.

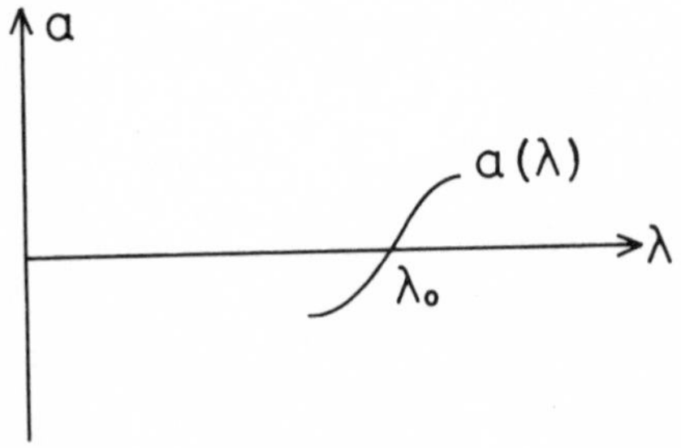


fig. 6.2

Unfortunately, there are many examples in the applications in which the situation depicted in fig. 6.2 is not to be expected. In fact, a very typical situation would be that depicted in fig. 6.3.

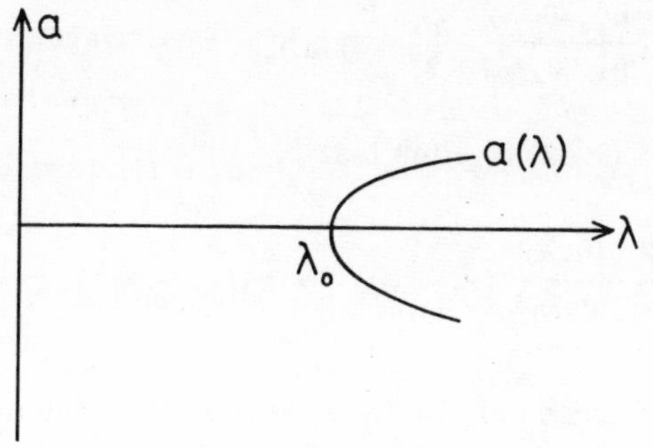


fig. 6.3

In this situation $b_a(0,\lambda_0) = 0$, although $b_\lambda(0,\lambda_0) \neq 0$. Thus the inequality (6.19) is violated although the inequality (6.20) still holds. The fact that $b_\lambda(0,\lambda_0) \neq 0$ implies that it is possible to solve for $\lambda$ as a function of $a$, i.e., $\lambda = \lambda(a)$ in a neighbourhood of $(0,\lambda_0)$. In this case the bifurcation equation becomes

$$(6.27) \qquad b(a,\lambda(a)) = 0.$$

If $\lambda$ is a differentiable function of $a$, we may write

$$(6.28) \qquad \lambda(a) = \lambda(0) + \lambda'(0)a + \frac{\lambda''(0)}{2}a^2 + 0(a^2).$$

Equ. (6.27) yields

$$(6.29) \qquad b_a(a,\lambda(a)) + b_\lambda(a,\lambda(a))\lambda'(a) = 0$$

or since by assumption $b_a(0,\lambda_0) = 0$,

$$(6.30) \qquad \lambda'(0) = 0.$$

After differentiating (6.27) a second time, it is found that $\lambda''(0)$ satisfies

$$(6.31) \qquad \lambda''(0) = -b_{aa}(0,\lambda_0)/b_\lambda(0,\lambda_0)$$

so that $\lambda(a)$ can be written

$$(6.32) \qquad \lambda(a) = \lambda_0 - \frac{b_{aa}(0,\lambda_0)}{b_\lambda(0,\lambda_0)} \frac{a^2}{2} + 0(a^2).$$

Thus if $b_{aa}(0,\lambda_0) \neq 0$ (6.32) implies that

$$(6.33) \qquad a = \pm\sqrt{\frac{2b_\lambda(0,\lambda_0)}{b_{aa}(0,\lambda_0)}(\lambda_0 - \lambda)} + 0((\lambda_0 - \lambda)^{\frac{1}{2}}).$$

The function $a(\lambda)$ is defined only in the case $\lambda_0 - \lambda$ has the same sign as $b_\lambda(0,\lambda_0)/b_{aa}(0,\lambda_0)$. Thus if $b_{aa}(0,\lambda_0) \neq 0$ the branching diagram has either the appearance in fig. 6.4a or 6.4b, depending on the sign of $b_\lambda(0,\lambda_0)/b_{aa}(0,\lambda_0)$.

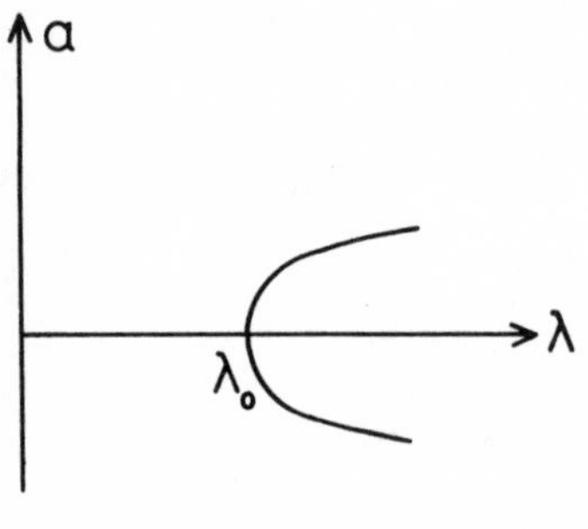


fig. 6.4a

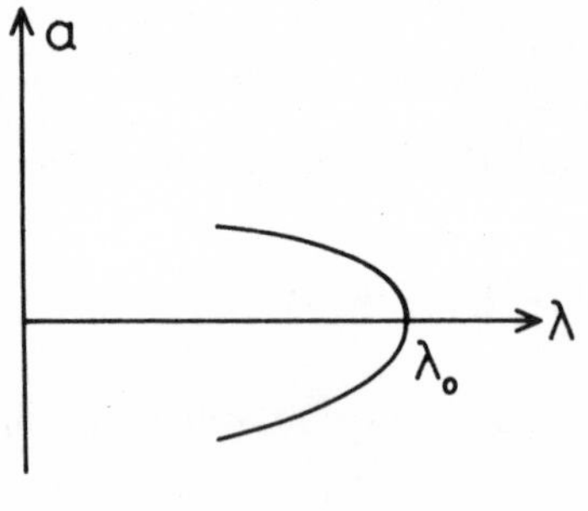


fig. 6.4b

Theorem (6.6): Assume the hypotheses of theorem (6.5) are satisfied except that

$$\int_{t_1}^{t_2} f_{uu}(u_0(t,\lambda_0),t,\lambda_0)\,\phi^3(t)\,dt = 0$$

and assume $b_{aa}(0,\lambda_0)$ exists and does not vanish. The boundary value problem (6.1) has two real solutions which are distinct from each other and from $u_0(t,\lambda_0)$ for $\lambda = \lambda_0$. These solutions are continuous in $\lambda$, identical at $\lambda = \lambda_0$, and are unique. The solutions are defined either for $0 < \lambda_0 - \lambda < \gamma$ or $0 < \lambda - \lambda_0 < \gamma$ but not both.

It is of interest to apply the above results to a special example. In particular, we consider the boundary value problem

(6.34a) $\quad (I(t)\,u_t)_t + \lambda g(u,t) = 0$

(6.34b) $\quad B_1 u = \alpha_1 u_t(t_1) + \beta_1 u(t_1) = 0$

(6.34c) $\quad B_2 u = \alpha_2 u_t(t_2) + \beta_2 u(t_2) = 0$

where $g$, $g_u$ and $g_{uu}$ are continuous and

(6.35) $\quad g(0,t) \equiv 0.$

In view of the above assumptions, the function $u_0(t,\lambda) \equiv 0$ is a solution to (6.34) for all $\lambda$. the variational problem corresponding to (6.34) is simply

(6.36a) $\quad (I(t)\,\dot{u}_t)_t + \lambda g_u(0,t)\,\dot{u} = 0$

(6.36b) $\quad B_1 u = 0$

(6.36c) $\quad B_2 \dot{u} = 0.$

The problem (6.36) is self-adjoint so that there exists a countably infinite set of eigenvalues $\lambda_0 < \lambda_1 < \lambda_2 < \ldots$ and corresponding eigenfunctions $\emptyset_n$. The eigenfunctions are unique if we require that

(6.37) $\quad \emptyset_n(t_1) = 1.$

Moreover, the eigenfunction $\emptyset_n$ has $n$ simple zeros in the open interval $t_1 < t < t_2$. In view of the above development it is clear that bifurcation is possible only in the case $\lambda = \lambda_n$. In addition, theorem (6.5) guarantees that bifurcation occurs at $\lambda = \lambda_n$ if

(6.38a) $$\lambda_n \int_{t_1}^{t_2} g_{uu}(0,t)\, \emptyset_n^3(t)\, dt \neq 0$$

(6.38b) $$\lambda_n \int_{t_1}^{t_2} g_u(0,t)\, \emptyset_n^2(t)\, dt \neq 0.$$

Assuming the above conditions are satisfied, the branching solutions can be written

$$u_n(t,\lambda) = a_n(\lambda)\, v_n(t,\lambda,a(\lambda))$$

since $u_0 \equiv 0$. At $\lambda = \lambda_n$ the function $v(t,\lambda_n,0) = \emptyset_n(t)$ and $\emptyset_n$ has n simple zeros. Therefore by continuity $u_n(t,\lambda)$ also has n simple zeros when $\lambda$ is sufficiently close to $\lambda_n$. One condition which guarantees that (6.38b) is satisfied is simply the condition that $g_u(0,t) > 0$. However, even in this case the condition (6.38a) may be violated. Even so, equ. (6.38b) implies that we can solve for $\lambda^n = \lambda^n(a)$. Thus we obtain a solution

(6.39) $$u_n(t,\lambda) = av(t,\lambda^n(a),a)$$

where

(6.40) $$\lambda^n(a) = \lambda_n + \lambda^n_{aa}(0)\frac{a^2}{2} + 0(a^2) \quad (\lambda_n = \lambda^n(0)).$$

It remains to determine $\lambda^n_{aa}(0)$. In order to determine this quantity, assume $u_a$ and $u_{aa}$ exist, which is justified if $g_{uu}$ exists. Differentiating (6.36), we find that $u_{aa}$ must satisfy the equation

(6.41) $$(Iu_{aat})_t + \lambda_n g_u(0,t)\, u_{aa} = \frac{\lambda_n}{3} g_{uuu}(0,t)\, \emptyset_n^3 - \lambda^n_{aa}(0)\, g_u(0,t)\, \emptyset_n(t)$$

at $(0,\lambda_n)$ and the boundary conditions

(6.42a) $$B_1 u_{aa} = 0$$

(6.42b) $$B_2 u_{aa} = 0.$$

Since by assumption the above problem has a non-trivial solution satisfying the boundary conditions (6.42), it follows that

$$\int_{t_1}^{t_2} [ - \frac{\lambda_n}{3} g_{uuu} (0,t) \phi_n^3 (t) - \lambda_{aa}^n (0) g_u (0,t) \phi_n (t) ] \phi_n (t) \, dt = 0$$

so that

$$\lambda_{aa}^{(n)} (0) = \frac{- \frac{\lambda_n}{3} \int_{t_1}^{t_2} g_{uuu} (0,t) \phi_n^4 (t) \, dt}{\int_{t_1}^{t_2} g_u (0,t) \phi_n^2 (t) \, dt}$$

Thus we see that $\lambda_{aa}^n (0) \neq 0$ if

$$(6.43) \qquad \lambda_n \int_{t_1}^{t_2} g_{uuu} (0,t) \phi_n^4 (t) \, dt \neq 0$$

and in this case theorem (6.6) applies, i.e., if (6.38b) and (6.43) are satisfied the problem (6.1) has two non-trivial solutions branching either to the right or to the left of $\lambda_n$.

The object now is to apply the theory developed above to the non-uniform, inextensible elastica. The equation describing the deformation of the elastica is

$$(6.44) \qquad (I(x) \psi_x)_x + \lambda \sin\psi = 0$$

where $\psi$ should satisfy the boundary conditions

$$(6.45) \qquad \psi_x (0) = \psi_x (1) = 0.$$

The function $\psi$ is the slope of the rod at a point $x$ (cf. fig. 6.5) and $I(x)$ is the modulus of elasticity. The first step in treating equ. (6.44) is to consider the variational problem

$$(6.46a) \qquad (I(x) \dot{\psi}_x)_x + \lambda (\cos \psi) \dot{\psi} = 0$$

$$(6.46b) \qquad \dot{\psi}_x (0) = \dot{\psi}_x (1) = 0$$

where $\psi \equiv 0$ is a solution for all $\lambda$. Thus the variational problem becomes

(6.47a) $(I(x)\dot{\psi}_x)_x + \lambda\dot{\psi} = 0$

(6.47b) $\dot{\psi}_x(0) = \dot{\psi}_x(1) = 0.$

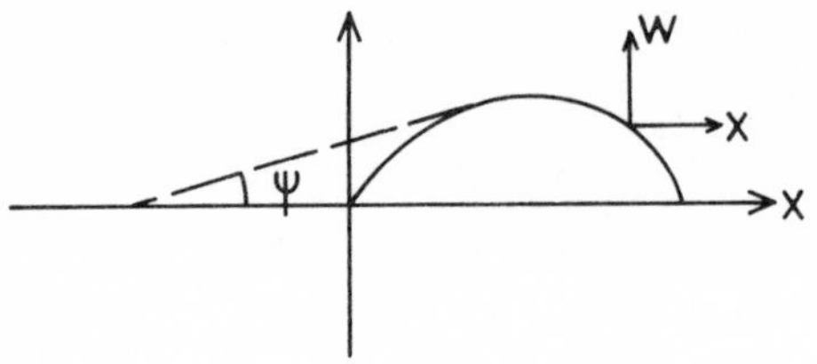


fig. 6.5

The smallest eigenvalue of the variational problem is $\lambda_0 = 0$ and the corresponding normalized eigenfunction is $\psi \equiv 1$. The rest of the eigenvalues are positive and can be ordered so that $0 < \lambda_1 < \lambda_2 < \ldots$; all of these eigenvalues are simple, i.e., correspond to exactly one linearly independent eigenfunction. For the eigenvalues $\lambda_n$ with $n \geq 1$, the integrals in (6.38a), (6.38b) and (6.43) become ($\psi \equiv 0$)

(6.48a) $$-\lambda_n \int_0^1 \sin\psi\, \phi_n^3\, dx = 0$$

(6.48b) $$\lambda_n \int_0^1 \cos\psi\, \phi_n^2\, dx = \lambda_n \int_0^1 \phi_n^2\, dx \neq 0$$

(6.48c) $$-\lambda_n \int_0^1 \cos\psi\, \phi_n^4\, dx = \lambda_n \int_0^1 \phi_n^4\, dx \neq 0.$$

It follows from the above development that branching occurs at each of the eigenvalues $\lambda_n$ with $n \geq 1$. In fact, it may be verified that locally these branching curves have the appearance depicted in fig. 6.6. These results may be summarised in the following theorem.

Theorem (6.7): For each $\lambda_n$ with $n \geq 1$ there exists a constant $\gamma_n$ such that for $\lambda_n \leq \lambda \leq \lambda_n + \gamma_n$ equ. (6.46a) with boundary conditions (6.46b) has two solutions $\psi_n(x,\lambda)$ and $-\psi_n(x,\lambda)$ with the properties that $\psi_n(x,\lambda_n) = 0$, $\psi_n(x,\lambda) \neq 0$ if $\lambda \neq \lambda_n$, and $\psi_n(x,\lambda)$ has $n-1$ simple zeros.

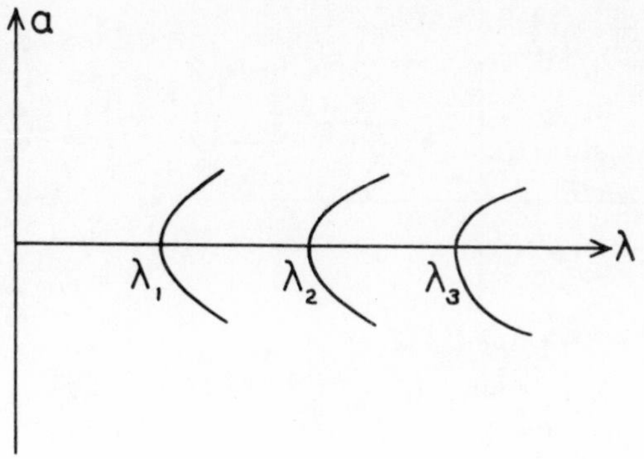


fig. 6.6

In the special case $I(x) \equiv 1$ it is possible to extend these results to obtain global results. Thus we consider the problem

(6.49a) $\psi_{xx} + \lambda \sin\psi = 0$

(6.49b) $\psi_x(0) = \psi_x(1) = 0.$

A direct linearization of this equation yields the problem

(6.50a) $\psi_{xx} + \lambda\psi = 0$

(6.50b) $\psi_x(0) = \psi_x(1) = 0.$

This latter linear problem has the solution

(6.51) $\psi_n = A_n \cos n\pi x, \quad n = 0,1,2,\ldots$

if $\lambda_n = (n\pi)^2$. For the moment we leave aside the case $\lambda_0 = 0$ and consider the problem (6.49) when $\lambda > 0$.

Theorem (6.8): The only solutions of problem (6.49) are $\psi = n\pi$ if $0 < \lambda < \lambda_1$.

Proof: Assume $\psi$ is a non-constant solution of (6.49). After multiplying equ. (6.49a) by $\psi$ and integrating by parts, it is found that the boundary conditions (6.49b) imply that

$$(6.52) \quad \int_0^1 (\psi_x^2 - \lambda\psi \sin\psi)\, dx = 0.$$

However, (6.52) yields

$$(6.53) \quad \int_0^1 (\psi_x^2 dx = \lambda \int_0^1 \psi \sin\psi\, d\psi \leq \lambda \int_0^1 \psi^2 dx$$

so that

$$(6.54) \quad \int_0^1 (\psi_x^2 - \lambda\psi^2)\, dx \leq 0.$$

On the other hand, as is well known

$$(6.55) \quad \min_{g \not\equiv \text{const}} \frac{\int_0^1 g_x^2 dx}{\int_0^1 g^2 dx} = \lambda_1 \leq \frac{\int_0^1 \psi_x^2}{\int_0^1 \psi^2 dx}.$$

Thus if $\lambda < \lambda_1$ (6.55) shows that

$$(6.56) \quad \int_0^1 (\psi_x^2 - \lambda\psi^2)\, dx > 0.$$

This contradiction proves the theorem. Q.E.D.

In the case of problem (6.49) it is possible to find an explicit expression for the curves which, as we already know, bifurcate from the eigenvalues $0 < \lambda_1 < \lambda_2 < \ldots.$ In order to determine these curves, multiply equ. (6.49a) by $\psi_x$. The resulting equation can be written

$$(6.57) \quad \frac{d}{dx}\left(\frac{1}{2}\psi_x^2 - \lambda\cos\psi\right) = 0.$$

Thus every solution of (6.49) satisfies the identity

(6.58) $$\psi_x^2 = 2\lambda(\cos\psi - \cos\alpha)$$

where $\psi(0) = \alpha$. Define a new variable $\emptyset$ such that

(6.59) $$\sin\frac{\psi}{2} = \ell \sin\emptyset$$

where

(6.60) $$\ell = \sin\alpha/2.$$

It follows from the identity

(6.61) $$\cos\psi = 1 - 2\sin^2\psi/2$$

that

(6.62) $$\emptyset_x = \mu\sqrt{(1 - \ell^2\sin^2\emptyset)}$$

with

(6.63) $$\mu = \sqrt{(\lambda)}.$$

In (6.62) the positive value of $\emptyset$ is chosen since the solutions will differ only in sign. It remains to determine the boundary conditions satisfied by $\emptyset$. From (6.59) it follows that

(6.64) $$\sin\psi(0)/2 = \sin\alpha/2 = \ell = \ell\sin\emptyset(0)$$

so that

(6.65) $$\sin\emptyset(0) = 1.$$

Thus

(6.66) $$\emptyset(0) = \emptyset_p = \frac{4p + 1}{2}\pi, \quad p = 0, \pm 1, \pm 2, \ldots.$$

In addition (6.50a) shows that

$$(6.67) \qquad \cos\psi(1) = \cos\alpha$$

or (cf. (6.61))

$$(6.68) \qquad \sin^2 \frac{\psi(1)}{2} = \sin^2 \alpha/2.$$

Equ. (6.68) combined with (6.59) yields

$$(6.69) \qquad \sin\phi(1) = \pm 1.$$

Therefore

$$(6.70) \qquad \phi(1) = \phi_q = \frac{2q+1}{2}\pi, \qquad q = \pm 1, \pm 2, \ldots.$$

The identity (6.62) implies that

$$(6.71) \qquad \mu x = \int_{\phi_p}^{\phi(x)} \frac{d\phi}{\sqrt{(1 - \ell^2 \sin\phi)}}$$

or evaluating this result at $x = 1$ we find

$$(6.72) \qquad \mu = \int_{\phi_p}^{\phi_q} \frac{d\phi}{\sqrt{(1 - \ell^2 \sin\phi)}}.$$

The function

$$(6.73) \qquad k(\ell) = \int_0^{\pi/2} \frac{d\phi}{\sqrt{(1 - \ell^2 \sin^2\phi)}}$$

is an elliptic integral of the first kind and it is easily verified that $k(\ell)$ has the properties: (i) $k(0) = \pi/2$, (ii) $dk/d\ell \gtrless 0$ when $\ell \gtrless 0$, and (iii) $k(\ell) \to \infty$ as $\ell \to \pm 1$. Since the interval of integration in the case of (6.72) is some multiple of $\pi$, equ. (6.72) can be written

$$(6.74) \qquad \mu = \mu_m(\ell) = 2mk(\ell), \qquad m = 1, 2, \ldots.$$

Note that $\mu_m(0) = m\pi = \lambda_m$. In view of the properties of $k(\ell)$, we obtain the diagram in fig. 6.7. If $\ell \ll 1$ the equ. (6.74) can be used to actually obtain a closed form expression for $\mu_m(\ell)$ which is valid in a neighbourhood of $\ell = 0$. In fact, assuming $\ell \ll 1$ equ. (6.73) implies that

$$(6.75) \qquad k(\ell) \approx \int_0^{\pi/2} (1 + \frac{\ell^2}{2} \sin^2 \phi)\, d\phi = \frac{\pi}{2}(1 + \frac{\ell^2}{4}).$$

so that

$$(6.76) \qquad \mu_m(\ell) \approx m\pi(1 + \ell^2/4).$$

Thus in a neighbourhood at $\ell = 0$, the function $\mu_m(\ell)$ is approximately a parabola.

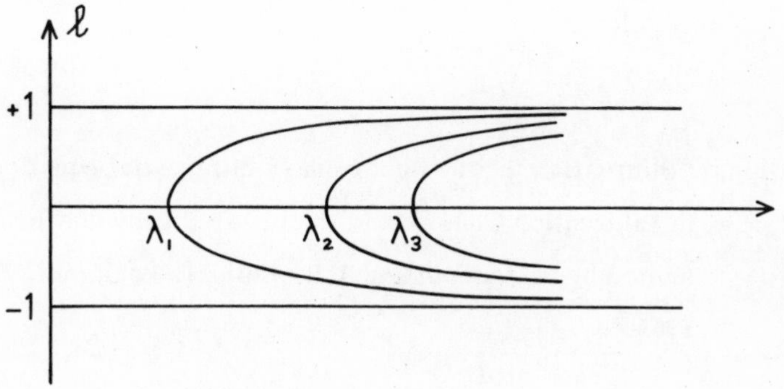


fig. 6.7

The above remarks give an essentially complete global description of the behaviour of the elastica when $\lambda \neq 0$. In the case $\lambda = 0$ the only solutions are $\psi = \psi_0 = \text{const}$. In fact, these solutions are of no physical interest since they correspond to the undeformed rod. To see this it suffices to note that the vertical and horizontal displacements of the rod $w$ and $u$ are defined in terms of $\psi$ by

$$(6.77a) \qquad w_x = \sin\psi$$

$$(6.77b) \qquad u_x = \cos\psi - 1.$$

Thus if one end of the elastica is fixed and the other end is free to slide on the $x$ axis, the functions $w$ and $u$ must satisfy the boundary conditions

(6.78) $\quad w(0) = w(1) = u(0) = 0.$

These boundary conditions in conjunction with (6.77a) yield

(6.79a) $\quad \sin\psi_0 = 0$

(6.79b) $\quad u(x) = (\cos\psi_0 - 1)x.$

Equ. (6.79a) implies that $\psi_0 = n\pi$. If in addition we require that at least one point of the rod lies to the right of the origin after deformation, then only even $n$ are possible, and hence $u(x) \equiv 0$, i.e., the elastica is undeformed.

## References


1 J.B. Keller, "Bifurcation theory for ordinary differential equations", article in Bifurcation Theory and Nonlinear Eigenvalue Problems (edited by S. Antman and J.B. Keller), Benjamin, New York, 1969.

2 E.L. Reiss, "Column buckling - an elementary example of bifurcation", article in Bifurcation Theory and Nonlinear Eigenvalue Problems (edited by S. Antman and J.B. Keller), Benjamin, New York, 1969.


## Other References


3 S.S. Antman, "The theory of rods", article in Handbuch Der Physik, Vol. VIa/2, Springer-Verlag, Berlin, 1972.

4 S.S. Antman, Kirchoff's Problem for Nonlinearly Elastic Rods. Quart. of Appl. Math. 32 (1974), 221-240.

Other References


5 S.S. Antman and K.B. Jordan, Qualitative Aspects of the Spatial Deformation of Nonlinearly Elastic Rods, Proc. of The Royal Society of Edinburgh (to appear).

6 M.G. Crandall and P.H. Rabinowitz, Nonlinear Sturm-Liouville Eigenvalue Problems and Topological Degree, Jour. of Math. and Mech. 19 (1970), 1083-1102.

7 R.W. Dickey, A Bifurcation Problem for a Fourth Order Nonlinear Ordinary Differential Equation, SIAM Jour. Appl. Math. 27 (1974), 93-101.

# 7 Buckling of the circular plate

Much of the interest in bifurcation theory has been stimulated by the question of the bifurcation points and post-buckling behaviour of the circular plate. This investigation has centred about the approximate model suggested by von Karman [1]. The von Karman equations for the rotationally symmetric buckled states can be written in the form

$$(7.1a) \qquad Gq + \lambda^2(1 - p)q = 0$$

$$(7.1b) \qquad Gp = -\frac{1}{2}q^2$$

where the differential operator $G$ is given by

$$(7.2) \qquad G = \frac{1}{r^3}\frac{d}{dr}\, r^3 \frac{d}{dr} \; .$$

The quantities $p$ and $q$ are related to the radial stress and slope of the plate and $\lambda^2$ is essentially the prescribed compressive force on the edge of the plate (cf. fig. 7.1)

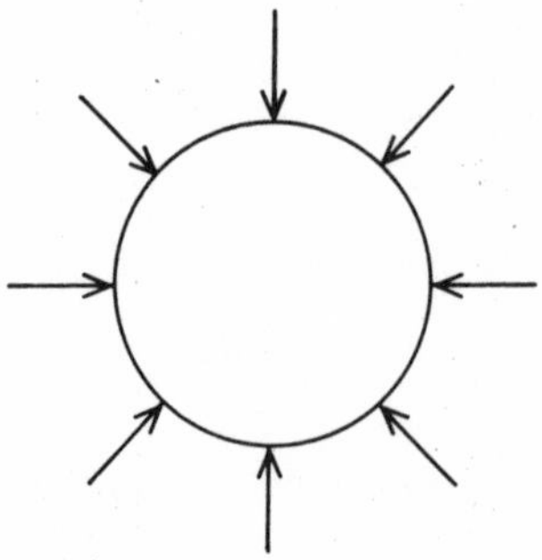

fig. 7.1

If the plate is clamped at the edge, the boundary conditions on (7.1) are

(7.3a) $q'(0) = p'(0) = 0$

(7.3b) $q(1) = p(1) = 0.$

On the other hand, if the edge is simply supported the boundary conditions become, in addition to (7.3a),

(7.4) $p(1) = q'(1) + (1 + \nu) q(1) = 0$

where $\nu$, the Poisson ratio, satisfies the inequality $0 < \nu < .5$.

In 1941 Friedrichs and Stoker [2] were able to show, using a variational approach, that the smallest eigenvalue of the linearization of (7.1) is a bifurcation point of (7.1). Moreover, they succeeded in obtaining results on the qualitative behaviour of the 'lowest buckled mode', including the asymptotic behaviour of this solution as $\lambda^2 \to \infty$. In 1962 Keller, Keller, and Reiss [3] were able to show, using methods similar to those described in Chapter 6, that each of the eigenvalues of the linearized version of (7.1) is a bifurcation point of (7.1). However, the results obtained in [3] are only valid in a neighbourhood of the eigenvalues. The extension to the global situation was obtained by Wolkowisky [4].

The linearization of equs. (7.1) leads to the single equation

(7.5) $G\bar{q} + \lambda^2 \bar{q} = 0.$

Equ. (7.5) with boundary conditions (7.3) has nontrivial solutions

(7.6) $\bar{q} = \bar{q}_n = \frac{C}{r} J_1(\lambda_n r)$

if $\lambda_n = j_{1,n}$ where $j_{1,n}$ is the n-th zero of the Bessel function $J_1$ and $C$ is an arbitrary constant. Wolkowisky was able to prove the following theorem:

Theorem (7.1): For all $\lambda > \lambda_n$ there exists at least $n$ pairs $(q_1, -q_1)$, $(q_2, -q_2), \ldots, (q_n, -q_n)$ of nontrivial solutions of (7.1) satisfying the boundary conditions (7.3). The function $q_j$ has $j-1$ zeros in the interval $0 < r < 1$.

In this chapter we will prove theorem (7.1) following the development in [4].

The proof of theorem (7.1) relies on the Schauder fixed point theorem (cf. [5]) which can be stated as follows:

Theorem (7.2): If $T$ is a completely continuous map taking a closed convex subset of a Banach space into itself, then $T$ has a fixed point.

Thus if $T$ satisfies the conditions of theorem (7.2), there exists an element $\phi$ such that $T(\phi) = \phi$. The object then is to define a transformation on a subset of an appropriate Banach space and verify that the conditions of the Schauder theorem are satisfied.

In the present situation the appropriate Banach space is the space $C_1$ of continuously differentiable functions $\phi$ on the interval $0 \leq r \leq 1$ with norm

$$\|\phi\| = \max_{0 \leq r \leq 1} |\phi| + \max_{0 \leq r \leq 1} |\phi'|.$$

In addition to $C_1$, we define the subset $S$ consisting of functions $\phi \in C_1$ with the properties

(7.7a) $\phi(0) = 1$

(7.7b) $\phi(1) = 0$

(7.7c) $-\dfrac{r\lambda^2}{2} \leq \phi'(r) \leq 0$

and the subset $S_\epsilon$ consisting of elements $\phi \in S$ such that

(7.8) $\phi \geq m_\epsilon(r)$

where

(7.9) $$m_\epsilon(r) = \begin{cases} \epsilon & 0 \leq r \leq 1 - \dfrac{1}{\lambda^2} \\ 0 & 1 - \dfrac{1}{\lambda^2} < r \leq 1 \end{cases} .$$

Lemma (7.1): If $\phi \in S$, then

(7.10a) $\quad \emptyset'(0) = 0$

(7.10b) $\quad 0 \le \emptyset(r) \le 1$

(7.10c) $\quad \|\emptyset\| \le 1 + \frac{\lambda^2}{2}$ .

Proof: Equ. (7.10a) follows from (7.7c) on setting $r = 0$. The inequality (7.10b) follows from (7.7a) and the fact that if $\emptyset \in S$ then $\emptyset$ is nonincreasing (cf. (7.7c)). The inequality (7.10c) follows from (7.7c) and (7.10b). Q.E.D

It is now convenient to define a mapping whose domain is $S_\epsilon$. Let $\emptyset \in S_\epsilon$ and let $C_j$ be the j-th eigenvalue and $q_j$ the corresponding eigenfunction of the linear Sturm-Liouville problem

(7.11a) $\quad (r^3 q_j')' + \lambda^2 r^3 (1 - C_j \emptyset) q_j = 0$

(7.11b) $\quad q_j'(0) = q_j(1) = 0$

It may be noted that the problem (7.11) has an infinite number of simple eigenvalues which satisfy the inequality

$$-\infty < C_k < C_{k-1} < \dots < C_1$$

for any integer $k$. The eigenfunction $q_j$ is of course determined only up to a multiplicative constant. However, this ambiguity may be removed by the normalizing condition

(7.11c) $\quad \int_0^1 \frac{1}{s^3} \int_0^s t^3 q_j^2(t)\, dt\, ds = 2$

The condition (7.11c) determines $q_j$ up to the sign. Finally, we define a function $\psi_j$ by

(7.11d) $\quad \psi_j = \frac{1}{2} \int_r^1 \frac{1}{s^3} \int_0^s t^3 q_j^2(t)\, dt\, ds.$

The mapping (7.11) can be written

$$(7.12) \qquad T_j(\emptyset) = \psi_j$$

where the integer $j$ implies that we are choosing the j-th eigenvalue $C_j$ of (7.11a). The object is to show that the mapping $T_j$ has a fixed point, say $P_j$ ($P_j \not\equiv 0$, cf. (7.11c)), if $\lambda > \lambda_n$ where $\lambda_n$ is the n-th eigenvalue of (7.5).

Lemma (7.2): If $\lambda > \lambda_n$ and $\emptyset \in S_\epsilon$, then

$$(7.13) \qquad C_j \geq (1 - \lambda_j^2/\lambda^2) > 0$$

when $j = 1,2,\ldots,n$.

Proof: Consider the Sturm-Liouville problem

$$(7.14a) \qquad (r^3 y')' + \lambda^2 r^3 (1 - \alpha) y = 0$$

$$(7.14b) \qquad y'(0) = y(1) = 0.$$

This problem has a nontrivial solution if (cf. (7.5))

$$(7.15) \qquad \lambda^2 (1 - \alpha_j) = \lambda_j^2$$

or equivalently if

$$(7.16) \qquad \alpha_j = 1 - \lambda_j^2/\lambda^2.$$

Since by assumption $\emptyset \in S_\epsilon$, the Sturm comparison theorem (cf. Chap. 1) and (7.10b) imply that $C_j \geq \alpha_j$. Q.E.D.

Lemma (7.3): If $\lambda > \lambda_1$ and $\emptyset \in S_\epsilon$, then

$$(7.17) \qquad C_1 \leq 1/\epsilon.$$

The eigenfunction $q_j$ has $j-1$ zeros in the interval $0 < r < 1$.

Proof: Consider the Sturm-Liouville problem

(7.18a) $(r^3 z_j')' + \lambda^2 r^3 (1 - \beta_j m_\epsilon(r)) z_j = 0$

(7.18b) $z_j'(0) = z_j(1) = 0.$

The eigenfunction $z_1$ corresponding to the largest eigenvalue $\lambda_1$ is determined only up to sign. Thus there is no loss of generality in assuming that $z_1(0) > 0$. The problem (7.18) can be solved explicitly in terms of Bessel functions and in fact may be verified that $z_1'(r) < 0$ for $1 - 1/\lambda^2 < r \leq 1$. It follows that $1 - \beta_1/\epsilon \geq 0$, i.e.,

(7.19) $\beta_1 \leq 1/\epsilon.$

On the other hand, since $\emptyset \in S_\epsilon$ the inequality (7.8) implies that $\emptyset \geq m_\epsilon(r)$. Thus the Sturm comparison theorem yields

(7.20) $C_1 \leq \beta_1 \leq 1/\epsilon$

i.e., $C_1$ is bounded above. In order to prove the second part of the lemma, it suffices to note that since $z_j(r)$ has $j-1$ zeros in the interval $0 < r < 1$ the Sturm comparison theorem implies that $q_j$ has $j-1$ zeros in this interval. Q.E.D.

In order to prove the complete continuity of the mapping $T_j$ we will need bounds on $\psi_j, q_j^2$, and $|r^3(q_j^2)'|$. These bounds are furnished by the following lemma:

Lemma (7.4): If $\lambda > \lambda_n$ and $\emptyset \in S_\epsilon$

(7.21) $q_j^2 \leq 4\lambda^2\psi_j \leq 4\lambda^2$

and

(7.22) $|r^2(q_j^2)'| \leq 2\lambda^4$

for $j = 1,2,\ldots,n.$

Proof: After multiplying (7.11a) by $q_j$ the result can be written

$$(7.23)\quad \frac{1}{2}\left[r^3 (q_j^{\,2})'\right]' + \lambda^2 r^3 q_j^{\,2} = r^3 (q_j')^2 + \lambda^2 r^3 C_j \emptyset q_j^{\,2}$$

or, since $C_j > 0$ (lemma (7.2)),

$$(7.24)\quad [r^3 (q_j^{\,2})']' + 2\lambda^2 r^3 q_j^{\,2} \geq 0$$

Equ. (7.24) implies that

$$(7.25)\quad (q_j^{\,2})' + \frac{2\lambda^2}{r^3} \int_0^r t^3 q_j^{\,2} dt \geq 0$$

which may be rewritten

$$(7.26)\quad \left[q_j^{\,2} + 2\lambda^2 \int_0^r \frac{1}{s^3} \left(\int_0^s t^3 q_j^{\,2} dt\right) ds\right]' \geq 0.$$

The inequality (7.26) implies that

$$q_j^{\,2} + 2\lambda^2 \int_0^r \frac{1}{s^3} \left(\int_0^s t^3 q_j^{\,2} dt\right) ds$$

is a monotone increasing function, so that

$$(7.27)\quad q_j^{\,2}(r) + 2\lambda^2 \int_0^r \frac{1}{s^3} \int_0^s t^3 q_j^{\,2} dt ds \leq 2\lambda^2 \int_0^1 \frac{1}{s^3} \int_0^s t^3 q_j^{\,2} dt ds$$

($q_j(1) = 0$ (cf. (7.11b))) or

$$(7.28)\quad q_j^{\,2}(r) \leq 2\lambda^2 \int_r^1 \frac{1}{s^3} \int_0^s t^3 q_j^{\,2} dt ds = 4\lambda^2 \psi_j(r) \leq 4\lambda^2 \psi_j(0).$$

Since $\psi_j(0) = 1$ (cf. (7.11c) and (7.11d)) we have proved (7.21). In order to prove (7.22) we note that (7.24) can be rewritten

$$(7.29)\quad \left[r^3 (q_j^{\,2})' + 2\lambda^2 \int_0^r t^3 q_j^{\,2} dt\right]' \geq 0$$

so that the function

$$r^3(q_j^2)' + 2\lambda^2 \int_0^r t^3 q_j^2 dt$$

is monotone increasing. Thus

$$(7.30) \qquad 0 \leq r^3(q_j^2)' + 2\lambda^2 \int_0^r t^3 q_j^2 dt \leq 2\lambda^2 \int_0^1 t^3 q_j^2 dt.$$

This can be rewritten

$$(7.31) \qquad -2\lambda^2 \int_0^r t^3 q_j^2 dt \leq r^3(q_j^2)' \leq 2\lambda^2 \int_r^1 t^3 q_j^2 dt$$

from which (7.22) follows. Q.E.D.

Theorem (7.3): If $\lambda > \lambda_n$ then $T_j$ is completely continuous on $S_\epsilon$ for $j = 1,2,\ldots,n$.

Proof: The first step is to show that $T_j$ is continuous on $S_\epsilon$, i.e., we wish to show that if $\{\emptyset_m\}$ is a sequence in $S_\epsilon$ such that $\emptyset_m \rightarrow \emptyset \in S_\epsilon$ in the $C_1$ norm, then $T_j(\emptyset_m) \rightarrow \psi_j$ where $T_j(\emptyset) = \psi_j$. Denote $T_j(\emptyset_m) = \psi_{jm}$. For each function $\emptyset_m$ there exists a corresponding eigenfunction $q_{jm}$ of (7.11a). If we require that $q_{jm}(0) > 0$ and that $q_{jm}$ satisfies the normalizing condition (7.11c), the eigenfunction is unique. Moreover, lemma (7.4) implies that $|q_{jm}|$ is bounded independent of m. The functions $q_{jm}$ satisfy the equation (cf. (7.11a))

$$(7.32) \qquad q_{jm} = -\lambda^2 \int_r^1 \frac{1}{s^3} \int_0^s t^3(1 - C_{jm}\emptyset_m) q_{jm} dt ds$$

where $C_{jm}$ is the j-th eigenvalue corresponding to $\emptyset_m$. Since lemma (7.3) implies that $0 < C_{jm} \leq 1/\epsilon$, it follows from (7.32) that $|q_{jm}'|$ is bounded independent of m. Thus the Arzela lemma shows the existence of a subsequence $\{q_{jm}\}$ which converges uniformly to a continuous function $q_j$. Since the eigenvalues of a Sturm-Liouville problem depend continuously on the coefficients (cf. [5]), it follows that $C_{jm}$ and hence $C_{jmk} \rightarrow C_j$ where $C_j$ is the j-th eigenvalue of

(7.33a) $(r^3 y')' + \lambda^2 r^3 (1 - C_j \emptyset) y = 0$

(7.33b) $y'(0) = y(1) = 0.$

It may now be shown that $q_j$ is a solution of (7.33). For this purpose write

$$(7.34) \quad |q_j + \lambda^2 \int_r^1 \frac{1}{s^3} \int_0^s t^3 (1 - C_j \emptyset) q_j \, dt\, ds| \leq |q_j - q_{jmk}|$$

$$+ \lambda^2 |\int_r^1 \frac{1}{s^3} \int_0^s t^3 \,[(1 - C_j \emptyset) q_j - (1 - C_{jmk} \emptyset_{mk}) q_{jmk}] dt\, ds|.$$

The right side of (7.34) can be made arbitrarily small by choosing $m_k$ sufficiently large. Actually we claim that the whole sequence $q_{jm} \to q_j$. This is seen as follows: If the whole sequence did not converge to $q_j$, then there must exist a second subsequence $q_{jm\ell}$ which converges to a different function $\tilde{q}_j$. However, by the same argument as above $\tilde{q}_j$ must satisfy (7.33), the normalizing condition (7.11c), and the condition $\tilde{q}_j(0) > 0$. Since (7.33) has a unique solution satisfying these conditions $\tilde{q}_j = q_j$, i.e., the sequence $q_{jm}$ converges. It is an immediate consequence of (7.11d) that $\psi_{jm} \to \psi_j$. This completes the proof of the continuity of $T_j$. In order to prove the complete continuity of $T_j$, it is necessary to show that $T_j(S_\epsilon)$ is compact, i.e., we need to show that every sequence in $T_j(S_\epsilon)$ has a convergent subsequence. However, this follows easily from the fact that if $\psi_j \in S_\epsilon$ then $|\psi_j|, |\psi_j'|$, and $|\psi_j''|$ are uniformly bounded (cf. (7.21) and (7.11d)) and hence if $\psi_{jm}$ is any subsequence in $T_j(S_\epsilon)$, then the Arzela lemma implies the existence of a subsequence which converges in the $C_1$ norm. Q.E.D.

Theorem (7.4): If $\lambda > \lambda_n$ there exists an $\epsilon^* > 0$ such that $T_j(S_{\epsilon^*}) \subseteq S_{\epsilon^*}$ for $j = 1, 2, \ldots, n$.

Proof: The proof is in two parts. First we show that if $\emptyset \in S_\epsilon$ (for any $\epsilon > 0$), then $T_j(\emptyset) = \psi_j \in S$ and secondly we show that there exists an $\epsilon^* > 0$ such that $T_j(S_{\epsilon^*}) \subset S_{\epsilon^*}$ The fact that $\psi_j$ is in $C_1$ follows from (7.11d).

Thus to show that $\psi_j \in S$ it is only necessary to verify that the conditions (7.7) are satisfied. However, (7.7a) and (7.7b) follow immediately from (7.11c) and (7.11d). In addition (7.11d) implies that

$$(7.35) \qquad \psi_j'(r) = -\frac{1}{2r^3}\int_0^r t^3 q_j^2 dt$$

and from (7.21)

$$(7.36) \qquad \psi_j'(r) \geq -r\lambda^2/2.$$

Assume there exists no $\epsilon > 0$ such that $T_j(S_\epsilon) \subset S_\epsilon$. This implies that for any sequence $\epsilon_m$ such that $\epsilon_m \to 0$ we could choose a function $\emptyset_m \in S_{\epsilon_m}$ such that $T_j(\emptyset_m) = \psi_{jm} \notin S_{\epsilon_m}$. Since $S$ is compact there exists a subsequence $\emptyset_{mk}$ such that $\emptyset_m \to \emptyset$ and $\psi_{jm_k} \to \psi_j$. The function $\psi_j \notin S_\epsilon$ for any $\epsilon > 0$, i.e., $\psi_j \notin \bigcup_{\epsilon > 0} S_\epsilon$, which implies $\psi_j \in S - \bigcup_{\epsilon > 0} S_\epsilon$. It follows that there exists at least one point $r^*$ such that $0 < r^* \leq 1 - 1/\lambda^2$ $(\psi_j(r^*) = 0$. On the other hand $\psi_j$ is nonincreasing and $\psi_j(1) = 0$. Therefore $\psi_j(r) \equiv 0$ for $1 - 1/\lambda^2 \leq r \leq 1$. Equivalently

$$(7.37) \qquad \lim_{m_k \to \infty} \int_{1-1/\lambda^2}^{1} \frac{1}{s^3}\int_0^s t^3 q_{jm_k}{}^2 dt\,ds = 0.$$

Lemma (7.4) implies that the derivative of the integrand of (7.37) is uniformly bounded so that (7.37) yields

$$(7.38) \qquad \lim_{m_k \to \infty} \frac{1}{s^3}\int_0^s t^3 q_{jm_k}{}^2 dt = 0 \quad (1 - 1/\lambda^2 \leq s \leq 1)$$

or at $s = 1$

$$(7.39) \qquad \lim_{m_k \to \infty} \int_0^1 t^3 q_{jm_k}{}^2 dt = 0.$$

Since the derivative of the integrand of (7.39) is bounded, (cf. (7.22)), we conclude that $\lim_{m_k \to \infty} q_{jm_k}{}^2 = 0$ for $0 \leq t \leq 1$ which contradicts the condition (7.11c).

This contradiction proves the existence of at least one value of $\epsilon = \epsilon^* > 0$ such that $T(S_{\epsilon^*}) \subseteq S_{\epsilon^*}$. Q.E.D.

Theorems (7.3) and (7.4) show that the mapping $T_j$ satisfies the conditions of the Schauder theorem. Thus we find that there exists an element $\tilde{p}_j \in S_{\epsilon^*}$ and that $T_j(\tilde{p}_j) = \tilde{p}_j$. Equivalently we have proved the existence of functions $\tilde{p}_j$ and $\tilde{q}_j$ such that

$$(7.40a) \quad (r^3\tilde{q}_j')' + \lambda^2 r^3 (1 - C_j \tilde{p}_j)\tilde{q}_j = 0$$

$$(7.40b) \quad \tilde{p}_j'(0) = \tilde{p}_j(1) = \tilde{q}_j'(0) = \tilde{q}_j(1) = 0$$

$$(7.40c) \quad (r^3 p_j')' = -\frac{r^3}{2} q_j^2.$$

In order to complete the proof of theorem (7.1) define

$$(7.41) \quad p_j = C_j \tilde{p}_j, \ q_j = \sqrt{(C_j)}\tilde{q}_j$$

(recall that $C_j > 0$). After making the change in variable (7.41), we find that $p_j$ and $q_j$ satisfy the equations (7.1) and boundary conditions (7.3).

The preceding discussion deals entirely with the clamped plate. However, these results can be extended to include the simply supported plate (cf. [4]).

## References


1 T. von Karman, Festigkeitsprobleme im Maschinenbau, Encyclopadie der Mathematischen Wissenschaften, 4, Leipzig, 1910, pp. 348-352.

2 K.O. Friedrichs and J.J. Stoker, The Nonlinear Boundary Value Problem of the Buckled Plate, Amer.J.Math. 63, 1941, pp. 839-888.

3 J.B. Keller, H.B. Keller and E.L. Reiss, Buckled States of Circular Plates, Q.Appl.Math. 20, 1962, pp. 55-65.

References

4 J.H. Wolkowisky, Existence of Buckled States of Circular Plates, Comm. on Pure and Appl. Math. 20, 1967, pp. 549-560.

5 R. Courant and D. Hilbert, Methods of Mathematical Physics, Vol. 1, Interscience Publishers, New York, 1953.

6 R. Courant and D. Hilbert, Methods of Mathematical Physics, Vol. 2, Interscience Publishers, New York, 1962.